Influenza: A Century of Research

Authored by

Irina Kiseleva

Department of Virology
Institute of Experimental Medicine
St. Petersburg
Russia

&

Natalie Larionova

Department of Virology
Institute of Experimental Medicine
St. Petersburg
Russia

Influenza: A Century of Research

Authors: Irina Kiseleva and Natalie Larionova

ISBN (Online): 978-1-68108-844-0

ISBN (Print): 978-1-68108-845-7

ISBN (Paperback): 978-1-68108-846-4

Acknowledgement:

In the memory of Dr. Alexander Klimov, an eminent virologist, who did so much for the development of influenza research, we used his original drawing for the cover of this book, with permission of his widow, Dr. Marina Khristova.

need for a court order if at any point you breach any terms of this License Agreement. In no event will any delay or failure by Bentham Science Publishers in enforcing your compliance with this License Agreement constitute a waiver of any of its rights.

3. You acknowledge that you have read this License Agreement, and agree to be bound by its terms and conditions. To the extent that any other terms and conditions presented on any website of Bentham Science Publishers conflict with, or are inconsistent with, the terms and conditions set out in this License Agreement, you acknowledge that the terms and conditions set out in this License Agreement shall prevail.

Bentham Science Publishers Ltd.
Executive Suite Y - 2
PO Box 7917, Saif Zone
Sharjah, U.A.E.
Email: subscriptions@benthamscience.net

BENTHAM SCIENCE

CONTENTS

FOREWORD

Almost 100 years have passed since the discovery of the first influenza virus. Intensive research and advances in scientific methods have largely filled the information gap and blank spots in virology, but not fully. "Influenza: A Century of Research" is a comprehensive Book covering in condensed form a success achieved over a century of studying the virus from the discovery of the first influenza virus through to the main aspects of influenza and the development of vaccines and antivirals.

Clearly written by two experts in the field, the book is easy to read. Illustrations in full color are simple for understanding and suggestive and the bibliography is quite comprehensive – the Book is filled with 855 references from the era of discovery of the first viruses until the last findings in the influenza research. The Book summarizes and systematizes available information on this regard and aims to contribute to emphasize the need for better control for influenza.

The book is divided into seven key chapters covering historical aspects, models, and substrates for studying influenza virus, replication, ecology and evolution of influenza virus, prophylaxis and treatment, the impact of influenza and COVID-19 pandemics on the circulation of seasonal respiratory viruses. As an expert in influenza chemotherapy, I am most impressed by the sections examined types of influenza vaccines and anti-influenza drugs. "Influenza: A Century of Research" offers a high level of consistency of chapters. Their flow constitutes a continuous narrative coupled with the text's clear, enables readers from professionals to students to grasp key principles of influenza virology with ease.

The discussion of the reasons contributed to insufficient control for influenza is of particular interest. We cannot predict the future but we can try to protect ourselves from the influenza burden. The contents of this book are intended for all those working or studying in the area of influenza. I have read this Book with great interest and can recommend it to a wide range of readers, who will have to evaluate how the authors coped with their task.

Vladimir V. Zarubaev,
St. Petersburg Pasteur Institute of Epidemiology and Microbiology,
St. Petersburg, Russia

PREFACE

Influenza is one of the most mysterious and ancient diseases that have been accompanying our civilization for centuries. The first written description of influenza-like disease belongs to Hippocrates (5th century BC). However, the exact age of the influenza virus is unknown.

Humankind has successfully defeated many dangerous infections that caused harm to human health and safety. Plaque, polio, smallpox, tuberculosis, measles, and many other infectious diseases were completely eradicated or controlled. In contrast, influenza control remains a serious problem for public health.

According to the World Health Organization, influenza is responsible for 3-5 million severe illnesses worldwide and up to 650 thousand respiratory deaths. The unpredictable character of the influenza virus spread is a significant threat to humans. In the last eight decades after the first influenza virus was isolated, huge progress has been made in studying, preventing, and treating influenza. Several influenza vaccines and antiviral compounds have been licensed. However, the influenza virus is still ahead of us. Therefore, this eBook aims to show how influenza virology has developed historically. We briefly demonstrate the tremendous success that has been achieved in the study of influenza over the past 100 years and discuss if there is a way to control this infection.

A lot has been written about influenza already. PubMed and other online resources supporting the search and retrieval of peer-reviewed biomedical and life sciences literature comprise over 200,000 articles, reviews, and books. In this eBook, we have tried to illuminate knowledge on influenza from a historical perspective, chronologically, with an emphasis on the virological part of the studies.

Chapter 1 describes a history of isolation of the first influenza viruses and dwells on the historical aspect of the development of appropriate substrates and models for studying influenza. Chapter 2 addresses the genome and capsid structure, molecular mechanisms of replication, and life cycle of the influenza virus. Mechanisms involved in the attenuation and virulence of the influenza virus are also discussed. Chapter 3 presents a comprehensive review of the influenza virus ecology and evolution. The origin of epidemic and pandemic influenza viruses is discussed as well. Chapter 4 focuses on influenza prophylaxis and treatment. Historical aspects of current achievements in this field were reviewed. Reasons contributed to insufficient control for influenza are highlighted in Chapter 5. The main properties of the influenza virus that may influence control for influenza are described. The COVID-19 pandemic affected both all aspects of our lives and the circulation of seasonal respiratory viruses. Chapter 6 describes some issues arising with the spread of pandemic viruses in general and SARS-CoV-2 infection in particular.

The short final Chapter 7 has been written in the form of conclusion; it is devoted to briefly summarize the 100-year study of influenza. Despite the huge number of research and years or rather centuries of research, a cure for the virus has not yet been found. The possibility of defeating influenza in the nearest future is discussed.

The intended audience for this eBook includes students of biological and medical colleges, Ph.D. students, post-docs, a wide range of virologists who are specialized in the field of influenza, and everyone interested in this infection.

CONSENT FOR PUBLICATION

Not applicable.

CONFLICT OF INTEREST

The author declares no conflict of interest, financial or otherwise.

ACKNOWLEDGEMENTS

Finally, the authors would like to thank Dr. Vladimir Zarubaev for writing the foreword, Bentham Science Publishers for the continuous support throughout the process of writing this book, and the Russian Science Foundation (grant 21-75-30003) for the financial support.

Irina Kiseleva

&

Natalie Larionova
Institute of Experimental Medicine
St. Petersburg
Russia

CHAPTER 1

Historical Aspects

Abstract: The first animal influenza A virus was isolated in 1931 by Richard Shope. The virus caused a highly contagious, influenza-like disease in pigs. Two years later, in 1933, the first human influenza A virus was isolated by Wilson Smith and colleagues. Soon after, in 1940, a representative of influenza virus type B was discovered by Thomas Francis, Jr. Being obligate intracellular parasites, viruses can be cultivated only within sensitive substrates. Three main substrates for the cultivation of influenza viruses are known: sensitive animals, embryonated chicken eggs, and tissue cultures. Today, in the twenties of the 21st century, sensitive animals are not often used for the isolation of the infectious virus. However, they are widely used to study and model a number of infectious diseases, including influenza. A list of these animals used for influenza research is very long, starting from ferrets and mice and ended with exotic zebrafish.

Keywords: Animal models, Embryonated chicken eggs, First human influenza viruses, Influenza, Nomenclature, Tissue culture.

INTRODUCTION

Influenza is one of the most mysterious and ancient diseases known for centuries. The exact age of the influenza virus is unknown. Humankind has successfully eradicated many deadly infections like plaque, polio, *etc*. While acknowledging the measures of the World Health Organization (WHO), Centers for Disease Control and Prevention (CDC), and other specialized agencies responsible for international and national public health, have taken to control influenza, it remains a serious cross-border problem. According to WHO, influenza is responsible for million cases of severe illnesses worldwide and up to 650 thousand respiratory deaths [1]. The unpredictable character of the influenza virus spread is a significant threat to humans.

Isolation of the first influenza viruses resulted in a rapid flowering of influenza research. The right choice of adequate animal models and substrates for virus cultivation is very important. Therefore, in this chapter, the historical aspect of the development of appropriate substrates and models for studying influenza is discussed.

Irina Kiseleva and Natalie Larionova

THE FIRST INFLUENZA VIRUSES

Influenza viruses belong to the *Orthomyxoviridae* family. Members of this family, which include seven genera [2, 3], are characterized by similarities in the structure of the viral particle and the way of reproduction. These are enveloped RNA viruses with a single-stranded segmented minus-stranded (negative-stranded) genome [4, 5]. Looking back in history, it is remarkable that the first influenza virus was isolated in Italy 119 years ago but classified as "fowl plaque" or "fowl pox" virus. Fifty years later, it was found that this is one of the highly pathogenic avian influenza A viruses.

In 1931, Richard Shope isolated a filterable agent that caused a highly contagious, influenza-like disease in pigs [6 - 8]. Soon after, during the 1933 epidemic, the first human influenza A virus was isolated by Smith *et al.* [9]. After several unsuccessful attempts to infect different species such as guinea pigs, mice, snakes, and hedgehogs, it was found that only ferrets exhibited catarrhal, nasal, and temperature symptoms of respiratory disease similar to humans. In 1940, the first human influenza B virus was discovered [10], and in 1968, the new influenza A virus subtype H3N2 was identified in tissue culture [11]. In 1947, in the United States, type C influenza virus C/Taylor/1233/1947 was identified in a human with upper respiratory symptoms [12].

In the early 1950s, the type D influenza virus (Sendai strain) that typically affects rodents and is not pathogenic for humans was isolated. However, later it was discovered that the Sendai virus belongs to the family *Paramyxoviridae.* The real influenza D virus belongs to the recently characterized new genus in the *Orthomyxoviridae* family. This virus was originally detected in pigs in the US [13]; however, cattle are now believed to be the main reservoir of the type D influenza viruses [13 - 15].

The taxonomic division of influenza viruses is based on their antigenic characteristics. According to the antigenic specificity of the internal structural components of the virion: the nucleoprotein (NP) and the matrix protein (M1), they are subdivided into four genera A, B, C and D [2, 3]. Type C and D have 7 RNA segments and encode 9 proteins, while types A and B have 8 RNA segments. Viruses belonging to different genera do not have common antigens, differ in epidemiological features and the severity of the disease they cause. Genetic reassortment between viruses of different genera does not occur. Comparative characteristics of influenza viruses A, B, C and D are presented in Table **1**.

Table 1. Types and subtypes of influenza viruses.

Type	Subtype (Genetic Lineage)	Host Range (Reservoir)	No Genome Segments	Refs.
A	18 HA (H1–H18) and 11 NA (N1–N11) subtypes	Humans, pigs, aquatic birds, horses, bats	8	[16 - 18]
B	Victoria and Yamagata lineages	Humans, harbor seals	8	[19, 20]
C	N/A[1]	Humans, pigs, cattle	7	[21 - 24]
D	N/A	Cattle, pigs, goats, sheep, horses, camels	7	[3, 13, 14, 25 - 29]

[1]Not applicable.

Influenza A virus causes the most severe diseases and may infect a wide range of host species. There are 18 hemagglutinin (HA) subtypes and 11 neuraminidase (NA) subtypes, respectively [16]. Among 198 potential influenza A subtype combinations, 131 subtypes have been detected in wild nature [30]. Serotypes of HA, in turn, form two phylogenetic groups that differ significantly from each other, but within each group, the domain encoding the HA stalk is antigenically similar: group 1 includes hemagglutinins H1, H2, H5, H6, H8, H9, H11 – H13, H16 and H17, group 2 - hemagglutinins H3, H4, H7, H10, H14 and H15 [31]. Two subtypes of influenza A viruses are routinely circulating in humans: A (H1N1)pdm09 and A (H3N2). Influenza virus ecology and evolution will be discussed in detail in Chapter **3**.

Human influenza A viruses cause annual epidemics and occasionally pandemics. In addition, viruses of zoonotic origin sporadically infect humans, causing severe respiratory infections and high mortality. Most zoonotic viruses are incapable of sustained human-to-human transmission, but mutations or reassortment with human influenza viruses in extremely rare cases leads to the emergence of a new virus with pandemic potential that can be transmitted by airborne droplets [32]. Only three serotypes of HA (H1, H2, and H3) and two serotypes of NA (N1 and N2) were integral parts of the pandemics pathogens and circulated widely among humans during the era of influenza studies [33]. The annual (seasonal) epidemics due to influenza A viruses are caused by the high rate of antigenic mutations that allow the virus to escape the immune defenses. Influenza viruses of genera B and C are not subdivided into serotypes, and no pandemics occur with their participation [34].

Influenza B viruses cause seasonal epidemics with the typical pattern of influenza infection. They are subject to antigenic variability too, although the rate of appearance mutations is 3-5 times lower compared to influenza A viruses [34,

35]. Since the early 1980s, two branches (lines) of the influenza B virus have been circulating alternately and jointly, differing significantly in antigenic characteristics of HA and NA and other properties [19, 35]. Influenza B virus causes less severe diseases than influenza A virus but can cause outbreaks and the influenza C virus causes acute respiratory illness most commonly in infants and young children, usually only associated with mild upper respiratory illness.

Influenza D virus has been identified recently. Its primary reservoir is cattle. Sheep, goats, pigs, horses, and camels are also susceptible to infection [3, 17, 36]. It is still unknown whether the influenza D virus causes disease in humans, but 97% of people in contact with cattle in Florida have antibodies to the influenza D virus [29].

MODELS FOR STUDYING INFLUENZA

Laboratory animals are widely used to study and model a number of diseases of an infectious and non-infectious nature, as well as for isolation of infectious agents. Animals, especially vertebrates, have repeatedly been used throughout the history of medical and biological discoveries and to this day, play a key role in biomedical researches. The first references to animal studies could be found in the writings of Greek philosophers and physicians Aristotle and Erasistratus, who most likely were the first to perform experiments on living animals. Louis Pasteur was the pioneer in the use of animals to prove the infectious nature of some diseases, such as anthrax [37, 38] or rabies [39].

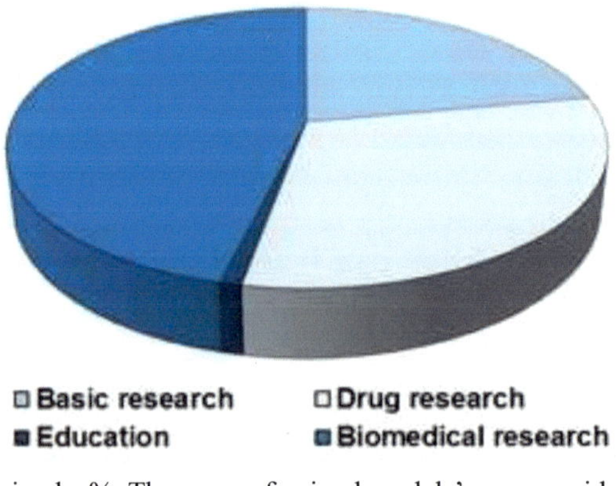

Fig. (1). Use of laboratory animals, %. The areas of animal models' use are wide.

Since the 18th century, the use of laboratory animals in experimentation became more common and the numbers of laboratory animal procedures conducted continue to rise progressively [40 - 42]. The areas of animal models' use can be seen in Fig. (**1**). The majority of animals are used in biomedical research (testing vaccines and biologicals, cancer research, heart diseases, and circulation research, *etc.*), basic research including military, space, *etc.*, drug research (toxicity tests, cosmetics, new antiviral compounds, *etc.*) and in education.

Animal research is now becoming even more prevalent - it is estimated that the number of mice and rats used in research has been increasing every year. Over 50 years ago, the principles of the 3Rs (Replacement, Reduction, and Refinement) providing a framework for performing more humane animal research were developed (Fig. **2**). 3Rs are guiding principles for the more ethical use of animals in preclinical studies. Replacement methods which avoid/replace the use of animals; reduction methods which minimize the number of animals used per experiment; refinement methods which minimize animal suffering and improve welfare. The questions of how to refine to lessen pain, reduce the number of animals used, and replace with non-animal methods and how many animals should be included in the group are under discussion [43, 44].

Fig. (2). Three guiding principles for the more ethical use of animals in preclinical studies (3Rs principles).

Ferrets were the first animals successfully used for the isolation of the first human influenza virus (see section "The first influenza viruses" above). Currently, ferrets are widely used to study the pathogenesis of influenza infection (Fig. **3**) [45 - 51] and transmissibility of influenza viruses [45, 47, 50]. Moreover, according to WHO recommendations [52], ferrets are used for preclinical characterization of potentially pandemic influenza vaccines [53 - 60].

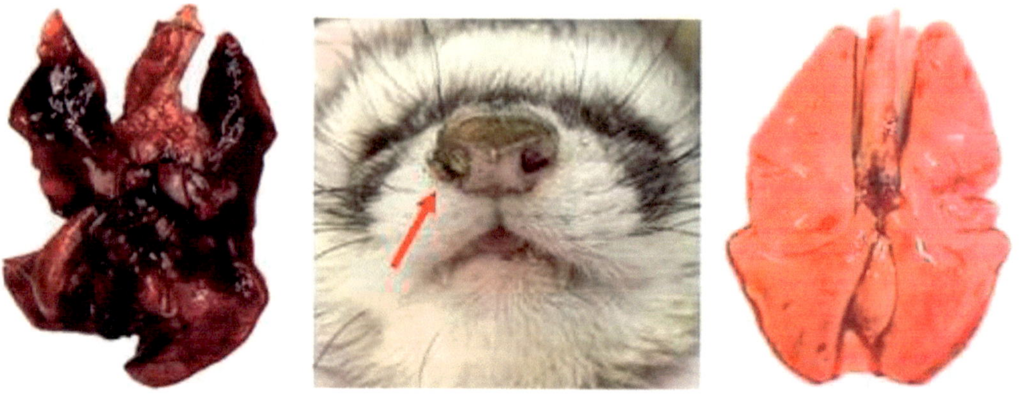

(a) (b) (c)

Fig. (3). Nasal symptoms and gross examination of the lungs of ferrets infected with A/South Africa/3626/2013 (H1N1)pdm09 influenza virus [61]. **(a, b)** The ferrets were inoculated with the virus. **(a)** Severe lung lesions. **(b)** Nasal discharge on the external nares (red arrow). **(c)** The lungs of the ferret inoculated with phosphate-buffered saline.

A wide range of other experimental animals (mice, poultry, guinea pigs, cotton rats, pigs, hamsters, macaques) are used to study the various aspects of the manifestation of influenza infection and the selection of therapeutic and prophylactic drugs to treat influenza infection (Table **2**) [47, 50, 62]. The most common models are mice (over 70% of all models used). A zebrafish has been actively used in the past few years to study the immune response to influenza and the selection of anti-influenza chemotherapy drugs [63 - 65]. Another exotic animal, the tree shrew, was shown to be physiologically and genetically related to primates, which make it a potential animal model of human diseases [66, 67]. Proper modeling of various aspects of the manifestation of influenza infection in the relevant sensitive animals is the key to scientific success.

Table 2. Animal models in influenza research.

Animals	Testing	Refs.
I. Standard models for testing influenza virus		
Ferrets	Influenza pathogenesis and therapy. Transmission. Preclinical evaluation of vaccine candidates and new therapeutics (antiviral effectiveness). Using polyclonal serum for immunological assays and development of LAIV candidates.	[47, 48, 50, 62, 68 - 71]
Mice	Influenza virus pathogenesis and immunity. Preclinical evaluation of vaccine candidates and new therapeutics (antiviral effectiveness).	[47, 62, 69, 70]
Non-human primates	Influenza virus pathogenesis and immunity.	[47, 62, 69, 70, 72 - 75]

(Table 2) cont.....

Animals	Testing	Refs.
Cotton rats	Influenza virus pathogenesis and immunity.	[47, 69, 70, 76 - 79]
II. Models occasionally used for testing influenza virus		
Chicken	Preclinical evaluation of potentially pandemic vaccine candidates of avian origin and veterinarian influenza vaccines.	[68, 69, 80 - 82]
Guinea pigs	Transmission.	[47, 69, 83, 84]
Hamsters	Influenza virus pathogenesis and immunity. Preclinical evaluation of vaccine candidates.	[47, 69, 85]
Pigs, minipigs	Preclinical evaluation of vaccine candidates against swine influenza.	[60, 62, 69, 86]
Tree shrews	Influenza virus pathogenesis and immunity. Transmission.	[66, 67]
Zebrafishes	Innate immunity. Preclinical evaluation of new therapeutics (antiviral effectiveness).	[63 - 65]
Cats	H5N1 influenza virus pathogenesis.	[87]
Dogs	H5N1 influenza virus pathogenesis. Preclinical evaluation of canine influenza vaccine candidates.	[88, 89]
Horses	Preclinical evaluation of equine influenza vaccine candidates.	[85, 90, 91]
Bats	H18N11 influenza virus pathogenesis and immunity.	[92]
Rabbits	Using polyclonal serum for determining antigen specificity and development of LAIV candidates.	[68, 93 - 95]
Rats	Using polyclonal serum for immunological assays and development of LAIV candidates.	[68, 96]
Sheep	Using polyclonal serum for immunological assays.	[97]
Goats	Using polyclonal serum for immunological assays.	[97]

The difference between a sensitive animal model as a substrate for virus isolation or research and a natural reservoir (host) of the influenza virus should be clearly understood (Fig. **4**). Mice, ferrets, guinea pigs, *etc.*, are very sensitive experimental animal models, but they never meet the influenza virus in their natural habitats. Between animals-substrates and animals-hosts, an additional group may be distinguished – animals who are not natural hosts but maybe occasionally infected. An accidental host accidentally harbors an organism that is not ordinarily parasitic in the particular species and usually does not infect it. Influenza A viruses can infect species other than the natural hosts in which they normally circulate on rare occasions.

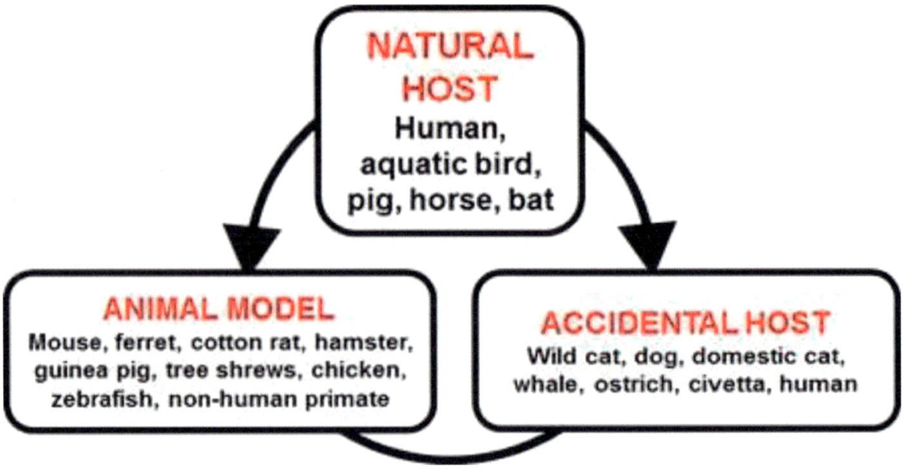

Fig. (4). Natural reservoirs and laboratory models for influenza A viruses.

For instance, humans could become accidental hosts for the avian influenza A (H5N1) virus. While avian influenza A viruses are highly species-specific, they may occasionally cross the species barrier to infect other species causing a disease of high lethality (Fig. **5**).

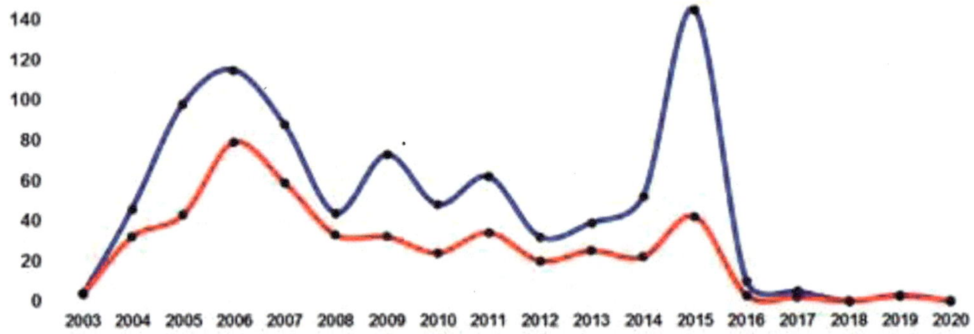

	2003	2004	2005	2006	2007	2008	2009	2010	2011	2012	2013	2014	2015	2016	2017	2018	2019	2020
Cases	4	46	98	115	88	44	73	48	62	32	39	52	145	10	4	0	1	0[2]
Deaths	4	32	43	79	59	33	32	24	24	20	25	22	42	3	2	0	1	0[2]
Death (%)[1]	100	70	44	69	67	75	44	50	55	63	64	42	29	30	50	0	100	0[2]

Fig. (5). Laboratory-confirmed human cases of avian influenza A (H5N1) reported to WHO in 2003-2020 and mortality rate (reprinted with permission from Book Publisher International) [98]. [1]The number of deaths (%) from the total number of laboratory-confirmed H5N1 cases. [2]Data from January 2020 through 28 February 2020.

This disease should not be confused with seasonal human influenza, generally caused by human H1N1pdm09, H3N2, or B viruses. Episodic transmission of avian influenza viruses to humans occurs when there is close contact with infected birds. Due to the high lethality and virulence, the avian influenza virus of subtype

H5N1 is one of the world's largest pandemic threats.

Ferrets being a highly experimental model that is sensitive to the influenza virus, could occasionally become the accidental hosts. During the influenza season, laboratory breeding ferrets can catch influenza from sick lab technicians or ferret's farmworkers. Influenza disease in ferrets can be fatal (Fig. **3**).

For obtaining adequate results and their correct assessment, three important conditions must be met:

- The right choice of anesthetic for animal study.
- Standardization of the infectious dose.
- The right choice of the virus as a model for research.

The Right Choice of Anesthetic for Animal Study

Virological and histopathological examinations occupy an important role in preclinical animal studies of influenza. A whole range of manipulations with animals is carried out under anesthesia to minimize their suffering. The refinement of anesthetic choice is an important part of experimental procedures. Ketamine is widely used in veterinary practice as a surgical anesthetic for general anesthesia. However, in some countries, ketamine is included in the list of narcotic drugs, psychotropic substances, and their precursors, which are subject to control, and its use is strictly limited. In veterinary practice, isoflurane is also broadly used as an anesthetic agent. Efficacy and safety of isoflurane have been shown in a number of nonclinical [99, 100] and clinical [101] trials. However, isoflurane may induce airway irritation [102] and should be used with caution in studies of the pathogenesis of respiratory viruses. It was shown that a five-fold inhalation of isoflurane might dramatically affect lung tissue and cause injuries in the form of hemorrhagic lung edema. Gross necropsy examination revealed gross lung lesions in a mock-inoculated group similar to animals infected with wild-type (WT) influenza virus [103]. This finding corresponded to a previous study [104]. The authors reported that general anesthesia with volatile agents, including isoflurane, provoked lung injury, acute inflammatory response, and leukocytic infiltration in rats. In contrast, the absence of any non-specific lung damages in ferrets caused by the injection of zoletil 100 was reported in studies of influenza virus [103, 105, 106] (Fig. **6**).

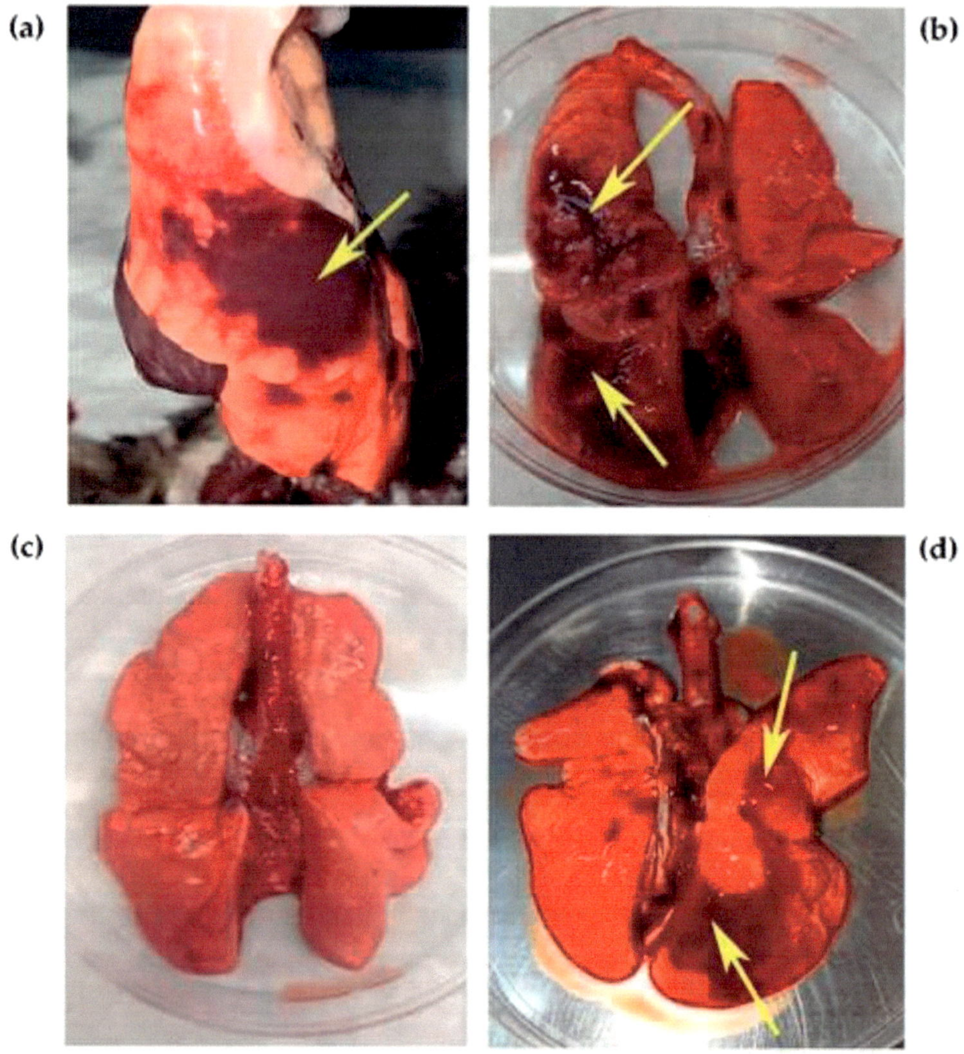

Fig. (6). Macroscopic view of the lungs of ferrets after multiple rounds of anesthesia (reprinted with permission from Vaccine Research) [103]. Yellow arrows – pulmonary hemorrhages. Ferrets received five applications of anesthesia. Isoflurane maintained inhalation anesthesia; intramuscular anesthesia was induced by injection with zoletil 100. At the end of the experiment, ferrets of all groups were humanely euthanized with a combination of Zoletil 100 and xylazine. **(a)** The ferret was anesthetized with isoflurane and received only PBS (placebo); **(b)** The ferret was anesthetized with isoflurane and inoculated with WT influenza B virus; **(c)** The ferret was anesthetized with zoletil 100 and received only PBS (placebo); **(d)** The ferret were anesthetized with zoletil 100 and inoculated with WT influenza B virus.

Standardization of the Infectious Dose

Standardization of the infectious dose is very important for study design and subsequent interpretation of the results. The literature describes three options for standardizing the dose of influenza virus to be used for infect experimental animals: (**i**) the 50% minimum infectious dose (MID_{50}) [107, 108]. Depending on the objectives, animals were usually infected with 10-100 MID_{50}. However, a preliminary determination of the MID_{50} is often a very expensive procedure, especially when it comes to animals such as monkeys or ferrets. As the option, the infectious dose to be inoculated to animals can be established (**ii**) by the measure of infectious virus titer in embryonated chicken eggs (50% embryo infectious dose, EID_{50}) or in tissue culture (50% tissue culture infectious dose, $TCID_{50}$; plaque-forming units, PFU); the animals are infected with one dose in the range of 5.0-7.0 $\log_{10}$ EID_{50} [95, 109 - 113] or 3.0-8.0 $\log_{10}$ $TCID_{50}$/PFU [105, 114 - 120]. (**iii**). If the virus to be tested is lung-adapted and may cause animals' death, the 50% lethal dose (LD_{50}) can be used [47, 107, 116].

The Right Choice of the Virus as a Model for Research

The susceptibility of the animal model to the influenza virus depends not only on the animal species but also on the virus strain to be tested in this model.

Historically, the pathogenesis of influenza infection has been studied in mice using lab lung-adapted (mouse-adapted) strains [47]. The number of classical mouse-adapted viruses is limited, for instance, A/Puerto Rico/8/34 (H1N1), B/Lee/40, A/Victoria/35/72 (H3N2), A/Aichi/2/68 (H3N2) and a few more. They have been used for pathogenicity studies for decades. However, today, in the twenties of the 21st century, all of them are antigenically distinct from circulating viruses.

The process of influenza virus adaptation to a new host is rather difficult. Fortunately, certain influenza viruses of the avian origin [108, 121], reconstructed 1918 Spanish influenza pandemic virus [122], and some H1N1pdm strains of the 2009 pandemic [123, 124] may cause disease in mice without prior adaptation. As it can be seen in Fig. (**7**), the non-mouse-adapted A/South Africa/3626/2013 (H1N1)pdm09 virus displayed severe lung lesions comparable to that of the mouse-adapted A/Puerto Rico/8/34 virus [61].

It allows making the overall conclusion that circulating non-mouse-adapted influenza viruses may be used for preclinical assessments of antiviral compounds or vaccine candidates. The advantage of non-adapted viruses is that (**i**) there is no need for additional lab work for their adaptation to model animals; (**ii**) no additional mutations occur in the genome of the virus during the long adaptation to another host.

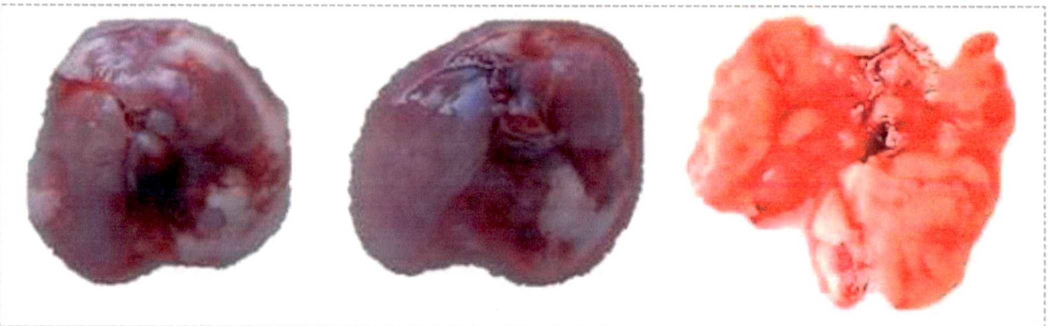

(a)	(b)	(c)
Mouse inoculated with 5.0 $\log_{10}$ EID_{50}/mouse of PR8[1], D14)	Mouse inoculated with 5.0 $\log_{10}$ EID_{50}/mouse of SA[2], D14)	Control mouse inoculated with PBS (D14)

Fig. (7). Gross examination of the lungs of mice infected with A/South Africa/3626/2013 (H1N1)pdm09 and A/Puerto Rico /8/34 (H1N1) viruses (adapted from [61]). [1]A/Puerto Rico/8/34 influenza virus. [2]A/South Africa/3626/2013 influenza virus.

SUBSTRATES FOR INFLUENZA VIRUS CULTIVATION

Being obligate intracellular parasites, viruses cannot be grown on any culture medium as bacteria can. Viruses can be cultivated only within suitable living hosts (substrates). Three main substrates for the cultivation of influenza viruses are known: (**i**) sensitive animal models; (**ii**) embryonated chicken eggs (herein simply referred to as "eggs"); (**iii**) tissue (cell) culture (Fig. **8**). The primary objectives of the cultivation of the influenza virus include isolation and identification of viruses in clinical specimens, the development of vaccines, and extensive research on influenza.

Experimental Animals

In the past, animals played an important role both as substrates for influenza virus isolation and as experimental models in influenza studies. As of today, in the twenties of the 21st century, the role of animals as a substrate became less prominent.

By the time the influenza virus was first isolated, laboratory animals were the only known substrate. In 1933, Smith *et al.* [9] isolated in ferrets the human influenza A strain. In 1937, another human influenza A virus was localized in mice by Smorodintsev *et al*. [125 - 127].

Embryonated Chicken Eggs

After chicken embryos were discovered as an alternative system for isolation of different viruses [128], a new era of virus isolation, including the influenza virus began.

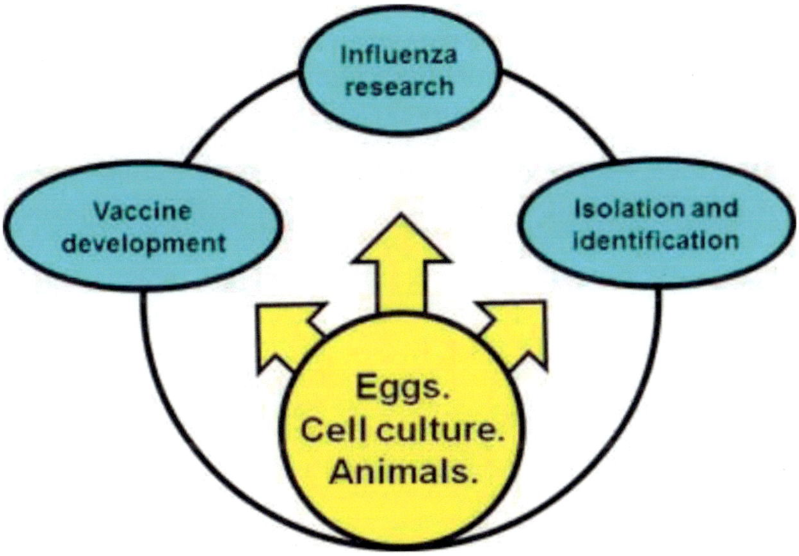

Fig. (8). Primary objectives of cultivation of influenza virus. Three main substrates for the cultivation of influenza viruses (sensitive animals, embryonated chicken eggs, and tissue culture) are known.

Tissue Culture

For decades, influenza viruses were propagated from clinical specimens in eggs or cultures of rhesus monkey kidney (Vero cell line) [129]. An alternative tissue culture system for the primary isolation and cultivation of influenza viruses was described in 1968 [130]. The authors found MDCK cells to be susceptible to a number of influenza virus strains.

In 1978, WHO officially suggested that continuous MDCK cells provide a satisfactory alternative to eggs and monkey kidney cultures for the isolation of influenza viruses from clinical specimens [131].

LABORATORY METHODS FOR DIAGNOSING INFLUENZA

Influenza can be difficult to diagnose through clinical signs alone because so many of the symptoms are similar to other acute respiratory diseases [132]. Laboratory methods are potentially more reliable and provide a much clearer picture of the nature of the disease. In 2011, a new edition of the "WHO manual for the laboratory diagnosis and virological surveillance of influenza" [133] was published, in which, on 153 pages, current methods used for the laboratory diagnosis of influenza are described in detail. In 2018 it was supplemented with "WHO information for the molecular detection of influenza viruses" [134] and the next year, WHO published revised guidance for National Influenza Centres which contained a description of the next-generation sequencing (NGS) of influenza viruses [135]. This document provides information on molecular detection and diagnostic protocols for the surveillance of influenza viruses in humans. NGS, a novel DNA sequencing technique, provides high speed and throughput that can be used in research and diagnostic virology [136].

The two main strategies in influenza diagnosis are to isolate the virus or its fragments directly from respiratory secretions (collection of specimens) or to measure the body's immune response to infection (retrospective analysis) (Fig. **9**). Laboratory diagnostic procedures for testing influenza include three groups of methods: (**i**) virus isolation by virological methods, (**ii**) serological diagnosis, and (**iii**) molecular identification (Fig. **9**).

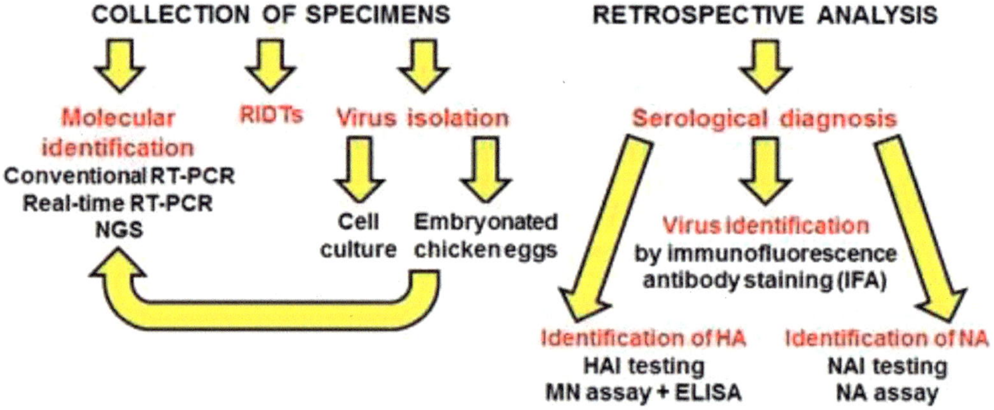

Fig. (9). Current strategies and laboratory procedures for diagnosing influenza.

Virus Culturing by Virological Methods

There are no dedicated methods in the diagnosis of influenza. The HA (hemagglutination) and HAI (inhibition of hemagglutination) tests described by Hirst [137 - 139] in the early 1940s of the XX century are successfully used to this day. The conventional gold standard for influenza diagnosis is cell-derived or egg-derived viral culture [140]. Viral culture tests are accurate and conclusive methods for diagnosing influenza and continue to provide a conclusive diagnosis. While the long turnaround times make this less useful in clinical settings, it is still a valid method for research. However, results can take 48-72 hours, up to 10 days.

Serological Diagnosis

Antibodies to influenza virus appear after 2 weeks and peak 4 to 7 weeks after infection. A variety of serological tests, including hemagglutination inhibition (HAI) assays and enzyme immunoassays (immunofluorescence antibody staining, IFA, enzyme-linked immunosorbent assay, ELISA), exist [140]. A ≥ 4-fold increase in influenza virus antibody titers in paired samples obtained at least 2 weeks apart establishes the serologic diagnosis of influenza. The requirement for paired samples obtained weeks apart renders this method not useful for urgent clinical diagnostic testing, but it can be helpful for retrospective analysis [133, 140].

Molecular Identification

Nowadays, viral culture has been somewhat superseded by tests based on reverse transcription polymerase chain reaction (RT-PCR), which produces results in hours rather than days and is considered the most sensitive and specific test for influenza [134, 135, 140]. Molecular tests afford the detection of influenza genetic material.

Rapid Influenza Diagnostic Tests (RIDTs)

RIDTs are immunoassays that can identify the presence of nucleoprotein of influenza virus in respiratory specimens, and display the result as positive *vs.* negative. RIDTs are the most common testing methods used to diagnose influenza [141, 142]. These point-of-care tests help in appropriate specimen collection and provide near-instantaneous results. However, not all RIDTs can distinguish between influenza A and B; and no RIDTs can distinguish between the varying strains of influenza A. Besides, if compared with other testing methods, RIDTs have the lowest sensitivities and the highest rate of false-negatives.

CONCLUDING REMARKS

The history of the detection of the influenza A virus dates back to the 1930s of the last century when the first swine influenza and human influenza viruses were isolated in animal models. From that moment, the number of animal models has grown substantially, and today is approaching 20 different species. Intensive study of sensitive substrates allowed us to discover many cell cultures highly sensitive to a variety of influenza viruses. Nevertheless, embryonated chicken eggs remain a widely used substrate for the isolation of influenza viruses and the production of influenza vaccines. Several hundred animal and human influenza viruses were isolated in embryonated chicken eggs and sensitive cell cultures, and their sequences were deposited in different databases. As of 16 March 2021, according to only one of the influenza databases, GISAID [143], a total of 1,448,836 sequences of 338,310 influenza virus strains of different types were deposited.

REFERENCES

[1] WHO. Media Centre. News release, 13 December 2017 [Available from: . https://www.who.int/mediacentre/news/statements/2017/flu/en/]

[2] ICTV. International Committee on Taxonomy of Viruses. 2019. [Available from: https://talk.ictvonline.org/ictv-reports/ictv_9th_report/negative-sense-rna-vir-ses-2011/w/negrna_viruses/209/orthomyxoviridae]

[3] Su S, Fu X, Li G, Kerlin F, Veit M. Novel Influenza D virus: Epidemiology, pathology, evolution and biological characteristics. Virulence 2017; 8(8): 1580-91. [http://dx.doi.org/10.1080/21505594.2017.1365216] [PMID: 28812422]

[4] Krammer F, Palese P. Influenza virus hemagglutinin stalk-based antibodies and vaccines. Curr Opin Virol 2013; 3(5): 521-30. [http://dx.doi.org/10.1016/j.coviro.2013.07.007] [PMID: 23978327]

[5] Nayak D, Shivakoti S, Balogun RA, Lee G, Zhou ZH. Structure, disassembly, assembly, and budding of influenza viruses.Textbook of influenza. 2nd ed. UK: Wiley Blackwell 2013; pp. 37-56. [http://dx.doi.org/10.1002/9781118636817.ch3]

[6] Van Epps HL. Influenza: exposing the true killer. J Exp Med 2006; 203(4): 803. [http://dx.doi.org/10.1084/jem.2034fta] [PMID: 16685764]

[7] Lewis PA, Shope RE. Swine influenza. II. A hemophilic bacillus from the respiratory tract of infected swine. J Exp Med 1931; 54(3): 361-71. [http://dx.doi.org/10.1084/jem.54.3.361] [PMID: 19869923]

[8] Shope RE. Swine influenza. III. Filtration experiments and etiology. J Exp Med 1931; 54(3): 373-85. [http://dx.doi.org/10.1084/jem.54.3.373] [PMID: 19869924]

[9] Smith W, Andrewes CH, Laidlaw PP. A virus obtained from influenza patients. Lancet 1933; 222(5732): 66-8. [http://dx.doi.org/10.1016/S0140-6736(00)78541-2]

[10] Francis T Jr. A new type of virus from epidemic influenza. Science 1940; 92(2392): 405-8. [http://dx.doi.org/10.1126/science.92.2392.405] [PMID: 17794275]

[11] Cockburn WC, Delon PJ, Ferreira W. Origin and progress of the 1968-69 Hong Kong influenza epidemic. Bull World Health Organ 1969; 41(3): 345-8. [PMID: 5309437]

[12] Taylor RM. Studies on survival of influenza virus between epidemics and antigenic variants of the virus. Am J Public Health Nations Health 1949; 39(2): 171-8. [http://dx.doi.org/10.2105/AJPH.39.2.171] [PMID: 18111593]

[13] Hause BM, Ducatez M, Collin EA, *et al.* Isolation of a novel swine influenza virus from Oklahoma in 2011 which is distantly related to human influenza C viruses. PLoS Pathog 2013; 9(2): e1003176. [http://dx.doi.org/10.1371/journal.ppat.1003176] [PMID: 23408893]

[14] Flynn O, Gallagher C, Mooney J, *et al.* Influenza D virus in cattle, Ireland. Emerg Infect Dis 2018; 24(2): 389-91. [http://dx.doi.org/10.3201/eid2402.170759] [PMID: 29350168]

[15] Hause BM, Collin EA, Liu R, *et al.* Characterization of a novel influenza virus in cattle and Swine: proposal for a new genus in the Orthomyxoviridae family. MBio 2014; 5(2): e00031-14. [http://dx.doi.org/10.1128/mBio.00031-14] [PMID: 24595369]

[16] Hutchinson EC. Influenza virus. Trends Microbiol 2018; 26(9): 809-10. [http://dx.doi.org/10.1016/j.tim.2018.05.013] [PMID: 29909041]

[17] Long JS, Mistry B, Haslam SM, Barclay WS. Host and viral determinants of influenza A virus species specificity. Nat Rev Microbiol 2019; 17(2): 67-81. [http://dx.doi.org/10.1038/s41579-018-0115-z] [PMID: 30487536]

[18] Wasik BR, Voorhees IEH, Parrish CR. Canine and feline influenza. Cold Spring Harb Perspect Med 2021; 11(1): a038562. [http://dx.doi.org/10.1101/cshperspect.a038562] [PMID: 31871238]

[19] Rota PA, Wallis TR, Harmon MW, Rota JS, Kendal AP, Nerome K. Cocirculation of two distinct evolutionary lineages of influenza type B virus since 1983. Virology 1990; 175(1): 59-68. [http://dx.doi.org/10.1016/0042-6822(90)90186-U] [PMID: 2309452]

[20] Osterhaus AD, Rimmelzwaan GF, Martina BE, Bestebroer TM, Fouchier RA. Influenza B virus in seals. Science 2000; 288(5468): 1051-3. [http://dx.doi.org/10.1126/science.288.5468.1051] [PMID: 10807575]

[21] Andrews BE, McDonald JC. Influenza virus C infection in England. BMJ 1955; 2(4946): 992-4. [http://dx.doi.org/10.1136/bmj.2.4946.992] [PMID: 13260642]

[22] Taylor RM. A further note on 1233 influenza C virus. Arch Gesamte Virusforsch 1951; 4(4): 485-500. [http://dx.doi.org/10.1007/BF01241168] [PMID: 13249484]

[23] Zhang H, Porter E, Lohman M, *et al.* Influenza C virus in cattle with respiratory disease, United States, 2016-2018. Emerg Infect Dis 2018; 24(10): 1926-9. [http://dx.doi.org/10.3201/eid2410.180589] [PMID: 30226175]

[24] To J, Torres J. Viroporins in the influenza virus. Cells 2019; 8(7): 654. [http://dx.doi.org/10.3390/cells8070654] [PMID: 31261944]

[25] Collin EA, Sheng Z, Lang Y, Ma W, Hause BM, Li F. Cocirculation of two distinct genetic and antigenic lineages of proposed influenza D virus in cattle. J Virol 2015; 89(2): 1036-42. [http://dx.doi.org/10.1128/JVI.02718-14] [PMID: 25355894]

[26] Dane H, Duffy C, Guelbenzu M, *et al.* Detection of influenza D virus in bovine respiratory disease samples, UK. Transbound Emerg Dis 2019; 66(5): 2184-7. [http://dx.doi.org/10.1111/tbed.13273] [PMID: 31228318]

[27] Song H, Qi J, Khedri Z, *et al.* An open receptor-binding cavity of hemagglutinin-esterase-fusion glycoprotein from newly-identified influenza D virus: Basis for its broad cell tropism. PLoS Pathog 2016; 12(1): e1005411. [http://dx.doi.org/10.1371/journal.ppat.1005411] [PMID: 26816272]

[28] Ferguson L, Olivier AK, Genova S, *et al.* Pathogenesis of influenza D virus in cattle. J Virol 2016; 90(12): 5636-42. [http://dx.doi.org/10.1128/JVI.03122-15] [PMID: 27030270]

[29] White SK, Ma W, McDaniel CJ, Gray GC, Lednicky JA. Serologic evidence of exposure to influenza D virus among persons with occupational contact with cattle. J Clin Virol 2016; 81: 31-3. [http://dx.doi.org/10.1016/j.jcv.2016.05.017] [PMID: 27294672]

[30] CDC. Types of influenza viruses. 2020. [Available from: https://www.cdc.gov/flu/about/viruses/types.html]

[31] Laursen NS, Wilson IA. Broadly neutralizing antibodies against influenza viruses. Antiviral Res 2013; 98(3): 476-83. [http://dx.doi.org/10.1016/j.antiviral.2013.03.021] [PMID: 23583287]

[32] Holmes EC. Evolution in health and medicine Sackler colloquium: The comparative genomics of viral emergence. Proc Natl Acad Sci USA. 107(Suppl 1): 1742-6. [http://dx.doi.org/10.1073/pnas.0906193106] [PMID: 19858482]

[33] Zambon M. Influenza and other emerging respiratory viruses. Viral Infections 2014; 42(1): 45-51. [http://dx.doi.org/10.1016/j.mpmed.2013.10.017] [PMID: 32288576]

[34] Hay AJ, Gregory V, Douglas AR, Lin YP. The evolution of human influenza viruses. Philos Trans R Soc Lond B Biol Sci 2001; 356(1416): 1861-70. [http://dx.doi.org/10.1098/rstb.2001.0999] [PMID: 11779385]

[35] Kanegae Y, Sugita S, Endo A, *et al.* Evolutionary pattern of the hemagglutinin gene of influenza B viruses isolated in Japan: cocirculating lineages in the same epidemic season. J Virol 1990; 64(6): 2860-5. [http://dx.doi.org/10.1128/jvi.64.6.2860-2865.1990] [PMID: 2335820]

[36] Asha K, Kumar B. Emerging influenza D virus threat: What we know so far! J Clin Med 2019; 8(2): E192. [http://dx.doi.org/10.3390/jcm8020192] [PMID: 30764577]

[37] Mock M, Fouet A. Anthrax. Annu Rev Microbiol 2001; 55: 647-71. [http://dx.doi.org/10.1146/annurev.micro.55.1.647] [PMID: 11544370]

[38] Pasteur L, Chamberland R. Summary report of the experiments conducted at Pouilly-le-Fort, near Melun, on the anthrax vaccination, 1881. Yale J Biol Med 2002; 75(1): 59-62. [PMID: 12074483]

[39] Pearce JM. Louis Pasteur and rabies: a brief note. J Neurol Neurosurg Psychiatry 2002; 73(1): 82. [http://dx.doi.org/10.1136/jnnp.73.1.82] [PMID: 12082056]

[40] Hudson M. Why do the numbers of laboratory animal procedures conducted continue to rise? An analysis of the Home Office Statistics of Scientific Procedures on Living Animals: Great Britain 2005. Altern Lab Anim 2007; 35(1): 177-87. [http://dx.doi.org/10.1177/026119290703500110] [PMID: 17411364]

[41] Taylor K, Alvarez LR. An estimate of the number of animals used for scientific purposes worldwide in 2015. Alternatives to laboratory animals. Altern Lab Anim 2019; 47(5-6): 196-213. [http://dx.doi.org/10.1177/0261192919899853] [PMID: 32090616]

[42] Taylor K, Gordon N, Langley G, Higgins W. Estimates for worldwide laboratory animal use in 2005. Altern Lab Anim 2008; 36(3): 327-42. [http://dx.doi.org/10.1177/026119290803600310] [PMID: 18662096]

[43] Baumans V. Science-based assessment of animal welfare: laboratory animals. Rev Sci Tech 2005; 24(2): 503-13. [http://dx.doi.org/10.20506/rst.24.2.1585] [PMID: 16358504]

[44] NC3Rs. What are the 3Rs? The National Centre for the Replacement, Reduction, & Refinement of Animals in Research. 2020. [Available from: https://www.nc3rs.org.uk/the-3rs].

[45] Belser JA, Eckert AM, Tumpey TM, Maines TR. Complexities in ferret influenza virus pathogenesis and transmission models. Microbiology and molecular biology reviews. Microbiol Mol Biol Rev 2016; 80(3): 733-44. [http://dx.doi.org/10.1128/MMBR.00022-16] [PMID: 27412880]

[46] Belser JA, Katz JM, Tumpey TM. The ferret as a model organism to study influenza A virus infection. Dis Model Mech 2011; 4(5): 575-9. [http://dx.doi.org/10.1242/dmm.007823] [PMID: 21810904]

[47] Bouvier NM, Lowen AC. Animal models for influenza virus pathogenesis and transmission. Viruses 2010; 2(8): 1530-63. [http://dx.doi.org/10.3390/v20801530] [PMID: 21442033]

[48] Enkirch T, von Messling V. Ferret models of viral pathogenesis. Virology 2015; 479-480: 259-70. [http://dx.doi.org/10.1016/j.virol.2015.03.017] [PMID: 25816764]

[49] Matsuoka Y, Lamirande EW, Subbarao K. The mouse model for influenza. In: R. Coico, T. Kowalik,

J. Quarles, B. Stevenson, R. Taylor, Eds. Current protocols in microbiology. 15G.3.1-15.G.3.30. The US – Wiley Interscience. 2009; Supplement 13.
[http://dx.doi.org/10.1002/9780471729259.mc15g03s13] [PMID: 25816764]

[50] Oh DY, Hurt AC. Using the ferret as an animal model for investigating influenza antiviral effectiveness. Front Microbiol 2016; 7: 80.
[http://dx.doi.org/10.3389/fmicb.2016.00080] [PMID: 26870031]

[51] Thangavel RR, Bouvier NM. Animal models for influenza virus pathogenesis, transmission, and immunology. J Immunol Methods 2014; 410: 60-79.
[http://dx.doi.org/10.1016/j.jim.2014.03.023] [PMID: 24709389]

[52] WHO. Biosafety risk assessment and guidelines for the production and quality control of human influenza pandemic vaccines. 2005. [Available from: http://www.who.int/biologicals/publications/trs/areas/vaccines/influenza/Annex%205%20human%20pandemic%20influenza.pdf].

[53] de Jong JC, Claas EC, Osterhaus AD, Webster RG, Lim WL. A pandemic warning? Nature 1997; 389(6651): 554.
[http://dx.doi.org/10.1038/39218] [PMID: 9335492]

[54] Isakova-Sivak I, de Jonge J, Smolonogina T, *et al.* Development and pre-clinical evaluation of two LAIV strains against potentially pandemic H2N2 influenza virus. PLoS One 2014; 9(7): e102339.
[http://dx.doi.org/10.1371/journal.pone.0102339] [PMID: 25058039]

[55] Major D, Chichester JA, Pathirana RD, *et al.* Intranasal vaccination with a plant-derived H5 HA vaccine protects mice and ferrets against highly pathogenic avian influenza virus challenge. Hum Vaccin Immunother 2015; 11(5): 1235-43.
[http://dx.doi.org/10.4161/21645515.2014.988554] [PMID: 25714901]

[56] Mann AJ, Noulin N, Catchpole A, *et al.* Intranasal H5N1 vaccines, adjuvanted with chitosan derivatives, protect ferrets against highly pathogenic influenza intranasal and intratracheal challenge. PLoS One 2014; 9(5): e93761.
[http://dx.doi.org/10.1371/journal.pone.0093761] [PMID: 24850536]

[57] Palker T, Kiseleva I, Johnston K, *et al.* Protective efficacy of intranasal cold-adapted influenza A/New Caledonia/20/99 (H1N1) vaccines comprised of egg- or cell culture-derived reassortants. Virus Res 2004; 105(2): 183-94.
[http://dx.doi.org/10.1016/j.virusres.2004.05.009] [PMID: 15351492]

[58] Rekstin A, Desheva Y, Kiseleva I, Ross T, Swayne D, Rudenko L. Live attenuated influenza H7N3 vaccine is safe, immunogenic and confers protection in animal models. Open Microbiol J 2014; 8: 154-62.
[http://dx.doi.org/10.2174/1874285801408010154] [PMID: 25685247]

[59] Stittelaar KJ, Veldhuis Kroeze EJ, Rudenko L, *et al.* Efficacy of live attenuated vaccines against 2009 pandemic H1N1 influenza in ferrets. Vaccine 2011; 29(49): 9265-70.
[http://dx.doi.org/10.1016/j.vaccine.2011.09.049] [PMID: 21945964]

[60] van der Laan JW, Herberts C, Lambkin-Williams R, Boyers A, Mann AJ, Oxford J. Animal models in influenza vaccine testing. Expert Rev Vaccines 2008; 7(6): 783-93.
[http://dx.doi.org/10.1586/14760584.7.6.783] [PMID: 18665776]

[61] Kiseleva I, Rekstin A, Al Farroukh M, *et al.* Non-mouse-adapted H1N1pdm09 virus as a model for influenza research. Viruses 2020; 12(6): E590.
[http://dx.doi.org/10.3390/v12060590] [PMID: 32485821]

[62] Barnard DL. Animal models for the study of influenza pathogenesis and therapy. Antiviral Res 2009; 82(2): A110-22.
[http://dx.doi.org/10.1016/j.antiviral.2008.12.014] [PMID: 19176218]

[63] Gabor KA, Goody MF, Mowel WK, *et al.* Influenza A virus infection in zebrafish recapitulates mammalian infection and sensitivity to anti-influenza drug treatment. Dis Model Mech 2014; 7(11):

1227-37.
[http://dx.doi.org/10.1242/dmm.014746] [PMID: 25190709]

[64] Goody MF, Sullivan C, Kim CH. Studying the immune response to human viral infections using zebrafish. Dev Comp Immunol 2014; 46(1): 84-95.
[http://dx.doi.org/10.1016/j.dci.2014.03.025] [PMID: 24718256]

[65] Sullivan C, Matty MA, Jurczyszak D, *et al.* Infectious disease models in zebrafish. Methods Cell Biol 2017; 138: 101-36.
[http://dx.doi.org/10.1016/bs.mcb.2016.10.005] [PMID: 28129840]

[66] Xu S, Li X, Yang J, *et al.* Comparative pathogenicity and transmissibility of pandemic H1N1, avian H5N1, and human H7N9 influenza viruses in tree shrews. Front Microbiol 2019; 10: 2955.
[http://dx.doi.org/10.3389/fmicb.2019.02955] [PMID: 31921093]

[67] Yuan B, Yang C, Xia X, *et al.* The tree shrew is a promising model for the study of influenza B virus infection. Virol J 2019; 16(1): 77.
[http://dx.doi.org/10.1186/s12985-019-1171-3] [PMID: 31174549]

[68] Kiseleva I, Larionova N, Fedorova E, Isakova-Sivak I, Rudenko L. New methodological approaches in the development of Russian live attenuated vaccine for pandemic influenza. Transl Biomed 2015; 6: 1-9.
[http://dx.doi.org/10.21767/2172-0479.100013]

[69] Margine I, Krammer F. Animal models for influenza viruses: implications for universal vaccine development. Pathogens 2014; 3(4): 845-74.
[http://dx.doi.org/10.3390/pathogens3040845] [PMID: 25436508]

[70] Haga T, Horimoto T. Animal models to study influenza virus pathogenesis and control. Open Antimicrob Agents J 2010; 2(2): 15-21.
[http://dx.doi.org/10.2174/1876518101002020015]

[71] Shcherbik S, Pearce N, Kiseleva I, *et al.* Implementation of new approaches for generating conventional reassortants for live attenuated influenza vaccine based on Russian master donor viruses. J Virol Methods 2016; 227: 33-9.
[http://dx.doi.org/10.1016/j.jviromet.2015.10.009] [PMID: 26519883]

[72] Kobasa D, Jones SM, Shinya K, *et al.* Aberrant innate immune response in lethal infection of macaques with the 1918 influenza virus. Nature 2007; 445(7125): 319-23.
[http://dx.doi.org/10.1038/nature05495] [PMID: 17230189]

[73] Morokutti A, Muster T, Ferko B. Intranasal vaccination with a replication-deficient influenza virus induces heterosubtypic neutralising mucosal IgA antibodies in humans. Vaccine 2014; 32(17): 1897-900.
[http://dx.doi.org/10.1016/j.vaccine.2014.02.009] [PMID: 24560674]

[74] Rimmelzwaan GF, Kuiken T, van Amerongen G, Bestebroer TM, Fouchier RA, Osterhaus AD. Pathogenesis of influenza A (H5N1) virus infection in a primate model. J Virol 2001; 75(14): 6687-91.
[http://dx.doi.org/10.1128/JVI.75.14.6687-6691.2001] [PMID: 11413336]

[75] Rimmelzwaan GF, Kuiken T, van Amerongen G, Bestebroer TM, Fouchier RA, Osterhaus AD. A primate model to study the pathogenesis of influenza A (H5N1) virus infection. Avian Dis 2003; 47(3) (Suppl.): 931-3.
[http://dx.doi.org/10.1637/0005-2086-47.s3.931] [PMID: 14575089]

[76] Eichelberger MC. The cotton rat as a model to study influenza pathogenesis and immunity. Viral Immunol 2007; 20(2): 243-9.
[http://dx.doi.org/10.1089/vim.2007.0017] [PMID: 17603841]

[77] Karlsson EA, Engel GA, Feeroz MM, *et al.* Influenza virus infection in nonhuman primates. Emerg Infect Dis 2012; 18(10): 1672-5.
[http://dx.doi.org/10.3201/eid1810.120214] [PMID: 23017256]

[78] Kuiken T, Rimmelzwaan GF, Van Amerongen G, Osterhaus AD. Pathology of human influenza A (H5N1) virus infection in cynomolgus macaques (Macaca fascicularis). Vet Pathol 2003; 40(3): 304-10.
[http://dx.doi.org/10.1354/vp.40-3-304] [PMID: 12724572]

[79] Ottolini MG, Blanco JCG, Eichelberger MC, *et al.* The cotton rat provides a useful small-animal model for the study of influenza virus pathogenesis. J Gen Virol 2005; 86(Pt 10): 2823-30.
[http://dx.doi.org/10.1099/vir.0.81145-0] [PMID: 16186238]

[80] Gambaryan AS, Lomakina NF, Boravleva EY, *et al.* Comparative safety, immunogenicity, and efficacy of several anti-H5N1 influenza experimental vaccines in a mouse and chicken models (Testing of killed and live H5 vaccine). Influenza Other Respir Viruses 2012; 6(3): 188-95.
[http://dx.doi.org/10.1111/j.1750-2659.2011.00291.x] [PMID: 21951678]

[81] Romanova J, Krenn BM, Wolschek M, *et al.* Preclinical evaluation of a replication-deficient intranasal DeltaNS1 H5N1 influenza vaccine. PLoS One 2009; 4(6): e5984.
[http://dx.doi.org/10.1371/journal.pone.0005984] [PMID: 19543385]

[82] Suguitan AL Jr, McAuliffe J, Mills KL, *et al.* Live, attenuated influenza A H5N1 candidate vaccines provide broad cross-protection in mice and ferrets. PLoS Med 2006; 3(9): e360.
[http://dx.doi.org/10.1371/journal.pmed.0030360] [PMID: 16968127]

[83] Kiseleva I, Larionova N, Bazhenova E, Dubrovina I, Rudenko L. Study of transmissibility of wild type and cold-adapted influenza viruses in guinea pigs. Influenza Other Respir Viruses 2011; 5 (Suppl. 1): 304-7.
[http://dx.doi.org/10.1111/j.1750-2659.2011.00209.x]

[84] Lowen AC, Mubareka S, Tumpey TM, García-Sastre A, Palese P. The guinea pig as a transmission model for human influenza viruses. Proc Natl Acad Sci USA 2006; 103(26): 9988-92.
[http://dx.doi.org/10.1073/pnas.0604157103] [PMID: 16785447]

[85] Daly JM, Yates RJ, Browse G, *et al.* Comparison of hamster and pony challenge models for evaluation of effect of antigenic drift on cross protection afforded by equine influenza vaccines. Equine Vet J 2003; 35(5): 458-62.
[http://dx.doi.org/10.2746/042516403775600433] [PMID: 12875323]

[86] Hansen AK, Farlov H, Bollen P. Microbiological monitoring of laboratory pigs. Lab Anim 1997; 31(3): 193-200.
[http://dx.doi.org/10.1258/002367797780596248] [PMID: 9230498]

[87] Rimmelzwaan GF, van Riel D, Baars M, *et al.* Influenza A virus (H5N1) infection in cats causes systemic disease with potential novel routes of virus spread within and between hosts. Am J Pathol 2006; 168(1): 176-83.
[http://dx.doi.org/10.2353/ajpath.2006.050466] [PMID: 16400021]

[88] Cleaveland S, Meslin FX, Breiman R. Dogs can play useful role as sentinel hosts for disease. Nature 2006; 440(7084): 605.
[http://dx.doi.org/10.1038/440605b] [PMID: 16572146]

[89] Lee DH, Bae SW, Park JK, *et al.* Virus-like particle vaccine protects against H3N2 canine influenza virus in dog. Vaccine 2013; 31(32): 3268-73.
[http://dx.doi.org/10.1016/j.vaccine.2013.05.023] [PMID: 23707159]

[90] Chambers TM, Quinlivan M, Sturgill T, *et al.* Influenza A viruses with truncated NS1 as modified live virus vaccines: pilot studies of safety and efficacy in horses. Equine Vet J 2009; 41(1): 87-92.
[http://dx.doi.org/10.2746/042516408X371937] [PMID: 19301588]

[91] Paillot R, Hannant D, Kydd JH, Daly JM. Vaccination against equine influenza: quid novi? Vaccine 2006; 24(19): 4047-61.
[http://dx.doi.org/10.1016/j.vaccine.2006.02.030] [PMID: 16545507]

[92] Gorka M, Schinköthe J, Ulrich R, *et al.* Characterization of experimental oro-nasal inoculation of

Seba's short-tailed bats (Carollia perspicillata) with bat influenza A virus H18N11. Viruses 2020; 12(2): E232.
[http://dx.doi.org/10.3390/v12020232] [PMID: 32093076]

[93] Klimov AI, Kiseleva IV, Desheva JA, Alexandrova GI, Cox NJ, Rudenko LG, Eds. Live attenuated reassortant influenza vaccine prepared using A/Leningrad/134/17/57 (H2N2) donor strain is genetically stable after replication in children 3–6 years of age. Options for the control of influenza IV; 2000; 951-954. Crete, Greece, 23rd-28th September 2000: Elsevier.

[94] Larionova N, Kiseleva I, Dubrovina I, Bazhenova E, Rudenko L. Peculiarities of reassortment of a cold-adapted influenza A master donor virus with influenza A viruses containing hemagglutinin and neuraminidase of avian H5N1 origin. Influenza Other Respir Viruses 2011; 5 (Suppl. 1): 346-9.
[http://dx.doi.org/10.1111/j.1750-2659.2011.00209.x]

[95] Larionova N, Kiseleva I, Isakova-Sivak I, *et al.* Live attenuated influenza vaccines against highlypathogenic H5N1 avian influenza: Development and preclinical characterization. J Vaccines Vaccin 2013; 4(8): 1000208.
[http://dx.doi.org/10.4172/2157-7560.1000208]

[96] Baibus DM, Konovalova NI, Prokopetz AV, Bil'danova ER, Eropkin M, Sominina AA. Potential use of rat polyclonal antisera for antigenic analysis of human influenza viruses. Epidermiol Vacc Prevent 2013; 1(68): 73-9. [Available from:https://www.researchgate.net/ publication/ 268815722_Potential_Use_of_Rat_Polyclonal_Antisera_for_Antigenic_Analysis_of_Hum an_Influenza_Viruses].

[97] NIBSC. Influenza reagents. 2020. [Available from: https://nibsc.org/products/brm_product_catalogue/influenza_reagents.aspx].

[98] Kiseleva I, Larionova N. Live attenuated potentially pandemic H5 vaccines on cold-adapted A/Leningrad/134/17/57 master donor virus backbone. In: Elrefaei AHA, Ed. Recent Progress in Microbiology and Biotechnology. BP International 2020; pp. 14-30. [Available from: https://www.researchgate.net/publication/343582633_Recent_Progress_in_Microbiology_and_Biotechnology_Vol_1].
[http://dx.doi.org/10.9734/bpi/rpmb/v1]

[99] Imai A, Steffey EP, Farver TB, Ilkiw JE. Assessment of isoflurane-induced anesthesia in ferrets and rats. Am J Vet Res 1999; 60(12): 1577-83.
[PMID: 10622172]

[100] Lichtenberger M, Ko J. Anesthesia and analgesia for small mammals and birds. Vet Clin North Am Exot Anim Pract 2007; 10(2): 293-315.
[http://dx.doi.org/10.1016/j.cvex.2006.12.002] [PMID: 17577554]

[101] Yasuda N, Lockhart SH, Eger EI II, *et al.* Kinetics of desflurane, isoflurane, and halothane in humans. Anesthesiology 1991; 74(3): 489-98.
[http://dx.doi.org/10.1097/00000542-199103000-00017] [PMID: 2001028]

[102] van Heerden PV, Wilson IH, Marshall FP, Cormack JR. Effect of humidification on inhalation induction with isoflurane. Br J Anaesth 1990; 64(2): 235-7.
[http://dx.doi.org/10.1093/bja/64.2.235] [PMID: 2317425]

[103] Kiseleva I, Stepanova E, Krutikova E, *et al.* Could trivalent LAIV protect against both genetic lineages of influenza B virus? Vaccine Rep 2020; 6(1): 12-23.
[http://dx.doi.org/10.29252/vacres.6.1.13]

[104] Nader-Djalal N, Knight PR, Bacon MF, Tait AR, Kennedy TP, Johnson KJ. Alterations in the course of acid-induced lung injury in rats after general anesthesia: volatile anesthetics *versus* ketamine. Anesth Analg 1998; 86(1): 141-6.
[http://dx.doi.org/10.1213/00000539-199801000-00029] [PMID: 9428869]

[105] Kim HM, Kang YM, Ku KB, *et al.* The severe pathogenicity of alveolar macrophage-depleted ferrets infected with 2009 pandemic H1N1 influenza virus. Virology 2013; 444(1-2): 394-403.

[http://dx.doi.org/10.1016/j.virol.2013.07.006] [PMID: 23890814]

[106] Lee DH, Kim JI, Lee JW, *et al.* Quantitative measurement of influenza virus replication using consecutive bronchoalveolar lavage in the lower respiratory tract of a ferret model. J Vet Sci 2014; 15(3): 439-42.
[http://dx.doi.org/10.4142/jvs.2014.15.3.439] [PMID: 24690606]

[107] Lu X, Edwards LE, Desheva JA, *et al.* Cross-protective immunity in mice induced by live-attenuated or inactivated vaccines against highly pathogenic influenza A (H5N1) viruses. Vaccine 2006; 24(44-46): 6588-93.
[http://dx.doi.org/10.1016/j.vaccine.2006.05.039] [PMID: 17030078]

[108] Lu X, Tumpey TM, Morken T, Zaki SR, Cox NJ, Katz JM. A mouse model for the evaluation of pathogenesis and immunity to influenza A (H5N1) viruses isolated from humans. J Virol 1999; 73(7): 5903-11.
[http://dx.doi.org/10.1128/JVI.73.7.5903-5911.1999] [PMID: 10364342]

[109] Klimov AI, Kiseleva IV, Alexandrova GI, Cox NJ, Eds. Genes coding for polymerase proteins are essential for attenuation of the cold–adapted A/Leningrad/134/17/57 (H2N2) influenza virus. Options for the Control of Influenza IV; 2001; 955-959. Crete, Greece, 23-28 September 2000: Elsevier.

[110] de Jonge J, Isakova-Sivak I, van Dijken H, *et al.* H7N9 live attenuated influenza vaccine is highly immunogenic, prevents virus replication, and protects against severe bronchopneumonia in ferrets. J Vaccines Vaccin 2016; 24(5): 991-1002.
[http://dx.doi.org/10.1038/mt.2016.23] [PMID: 26796670]

[111] Huang SSH, Banner D, Paquette SG, Leon AJ, Kelvin AA, Kelvin DJ. Pathogenic influenza B virus in the ferret model establishes lower respiratory tract infection. J Gen Virol 2014; 95(Pt 10): 2127-39.
[http://dx.doi.org/10.1099/vir.0.064352-0] [PMID: 24989173]

[112] Middleton D, Rockman S, Pearse M, *et al.* Evaluation of vaccines for H5N1 influenza virus in ferrets reveals the potential for protective single-shot immunization. J Virol 2009; 83(15): 7770-8.
[http://dx.doi.org/10.1128/JVI.00241-09] [PMID: 19457991]

[113] Paquette SG, Banner D, Huang SS, *et al.* Influenza transmission in the mother-infant dyad leads to severe disease, mammary gland infection, and pathogenesis by regulating host responses. PLoS Pathog 2015; 11(10): e1005173.
[http://dx.doi.org/10.1371/journal.ppat.1005173] [PMID: 26448646]

[114] Broadbent AJ, Santos CP, Anafu A, Wimmer E, Mueller S, Subbarao K. Evaluation of the attenuation, immunogenicity, and efficacy of a live virus vaccine generated by codon-pair bias de-optimization of the 2009 pandemic H1N1 influenza virus, in ferrets. Vaccine 2016; 34(4): 563-70.
[http://dx.doi.org/10.1016/j.vaccine.2015.11.054] [PMID: 26655630]

[115] Carolan LA, Rockman S, Borg K, *et al.* Characterization of the localized immune response in the respiratory tract of ferrets following infection with influenza A and B viruses. J Virol 2015; 90(6): 2838-48.
[http://dx.doi.org/10.1128/JVI.02797-15] [PMID: 26719259]

[116] Chen JR, Yu YH, Tseng YC, *et al.* Vaccination of monoglycosylated hemagglutinin induces cross-strain protection against influenza virus infections. Proc Natl Acad Sci USA 2014; 111(7): 2476-81.
[http://dx.doi.org/10.1073/pnas.1323954111] [PMID: 24469815]

[117] Pascua PN, Marathe BM, Burnham AJ, *et al.* Competitive fitness of influenza B viruses possessing E119A and H274Y neuraminidase inhibitor resistance-associated substitutions in ferrets. PLoS One 2016; 11(7): e0159847.
[http://dx.doi.org/10.1371/journal.pone.0159847] [PMID: 27466813]

[118] Russier M, Yang G, Marinova-Petkova A, *et al.* H1N1 influenza viruses varying widely in hemagglutinin stability transmit efficiently from swine to swine and to ferrets. PLoS Pathog 2017; 13(3): e1006276.
[http://dx.doi.org/10.1371/journal.ppat.1006276] [PMID: 28282440]

[119] Xu L, Bao L, Deng W, *et al.* The mouse and ferret models for studying the novel avian-origin human influenza A (H7N9) virus. Virol J 2013; 10: 253. [http://dx.doi.org/10.1186/1743-422X-10-253] [PMID: 23927489]

[120] Kim EH, Park SJ, Kwon HI, *et al.* Mouse adaptation of influenza B virus increases replication in the upper respiratory tract and results in droplet transmissibility in ferrets. Sci Rep 2015; 5: 15940. [http://dx.doi.org/10.1038/srep15940] [PMID: 26526113]

[121] Belser JA, Szretter KJ, Katz JM, Tumpey TM. Use of animal models to understand the pandemic potential of highly pathogenic avian influenza viruses. Adv Virus Res 2009; 73: 55-97. [http://dx.doi.org/10.1016/S0065-3527(09)73002-7] [PMID: 19695381]

[122] Tumpey TM, Basler CF, Aguilar PV, *et al.* Characterization of the reconstructed 1918 Spanish influenza pandemic virus. Science 2005; 310(5745): 77-80. [http://dx.doi.org/10.1126/science.1119392] [PMID: 16210530]

[123] Belser JA, Wadford DA, Pappas C, *et al.* Pathogenesis of pandemic influenza A (H1N1) and triple-reassortant swine influenza A (H1) viruses in mice. J Virol 2010; 84(9): 4194-203. [http://dx.doi.org/10.1128/JVI.02742-09] [PMID: 20181710]

[124] Itoh Y, Shinya K, Kiso M, *et al. In vitro* and *in vivo* characterization of new swine-origin H1N1 influenza viruses. Nature 2009; 460(7258): 1021-5. [http://dx.doi.org/10.1038/nature08260] [PMID: 19672242]

[125] Smorodintseff AA, Tushinsky MD, Drobyshevskaja AI, Korovin AA, Osetroff AI. Investigation on volunteers infected with influenza virus. Am J Med Sci 1937; 194: 159-70. [http://dx.doi.org/10.1097/00000441-193708000-00002]

[126] Zhdanov VM, Aleksandrova GI, Gendon IuZ. [Live influenza vaccine in the USSR: its development, study and practical use]. Zh Mikrobiol Epidemiol Immunobiol 1986; 7(7): 3-13. [PMID: 3529762]

[127] Zhilova GP, Alexandrova GI, Zykov MP, Smorodintsev AA. Some problems of modern influenza prophylaxis with live vaccine. J Infect Dis 1977; 135(4): 681-6. [http://dx.doi.org/10.1093/infdis/135.4.681] [PMID: 853250]

[128] Goodpasture EW, Woodruff AM, Buddingh GJ. The cultivation of vaccine and other viruses in the chorioallantoic membrane of chick embryos. Science 1931; 74(1919): 371-2. [http://dx.doi.org/10.1126/science.74.1919.371] [PMID: 17810781]

[129] Govorkova EA, Kaverin NV, Gubareva LV, Meignier B, Webster RG. Replication of influenza A viruses in a green monkey kidney continuous cell line (Vero). J Infect Dis 1995; 172(1): 250-3. [http://dx.doi.org/10.1093/infdis/172.1.250] [PMID: 7797924]

[130] Gaush CR, Smith TF. Replication and plaque assay of influenza virus in an established line of canine kidney cells. Appl Microbiol 1968; 16(4): 588-94. [http://dx.doi.org/10.1128/AM.16.4.588-594.1968] [PMID: 5647517]

[131] Davies HW, Appleyard G, Cunningham P, Pereira MS. The use of a continuous cell line for the isolation of influenza viruses. Bull World Health Organ 1978; 56(6): 991-3. [PMID: 310738]

[132] Warren-Gash C. Diagnosing influenza for research. Opticon 1826; 2012(14): 37-9. [http://dx.doi.org/10.5334/opt.al]

[133] WHO. Manual for the laboratory diagnosis and virological surveillance of influenza. 2011. [Available from: https://www.who.int/influenza/gisrs_laboratory/manual_diagnosis_surveillance_influenza/en/]

[134] WHO. Information for the molecular detection of influenza viruses. 2018. [Available from: https://www.who.int/influenza/gisrs_laboratory/Protocols_influenza_virus_detection_Nov_2018.pdf?ua=1]

[135] WHO. Next-generation sequencing of influenza viruses. 2019. [Available from: https://www.who.int/

influenza/gisrs_laboratory/national_influenza_centres/NGS_guidance_for_NICs.pdf?ua=1]

[136] Barzon L, Lavezzo E, Militello V, Toppo S, Palù G. Applications of next-generation sequencing technologies to diagnostic virology. Int J Mol Sci 2011; 12(11): 7861-84. [http://dx.doi.org/10.3390/ijms12117861] [PMID: 22174638]

[137] Hirst GK. The quantitative determination of influenza virus and antibodies by means of red cell agglutination. J Exp Med 1942; 75(1): 49-64. [http://dx.doi.org/10.1084/jem.75.1.49] [PMID: 19871167]

[138] Hirst GK. Adsorption of influenza hemagglutinins and virus by red blood cells. J Exp Med 1942; 76(2): 195-209. [http://dx.doi.org/10.1084/jem.76.2.195] [PMID: 19871229]

[139] Hirst GK. Studies of antigenic differences among strains of influenza A by means of red cell agglutination. J Exp Med 1943; 78(5): 407-23. [http://dx.doi.org/10.1084/jem.78.5.407] [PMID: 19871338]

[140] Kumar S, Henrickson KJ. Update on influenza diagnostics: lessons from the novel H1N1 influenza A pandemic. Clin Microbiol Rev 2012; 25(2): 344-61. [http://dx.doi.org/10.1128/CMR.05016-11] [PMID: 22491775]

[141] CDC. Rapid influenza diagnostic tests. 2021. [Available from: https://www.cdc.gov/flu/professionals/ diagnosis /clinician_guidance_ridt.htm]

[142] WHO. Influenza rapid diagnostic tests. 2010. [Available from: https://apps.who.int/iris/bitstream/ handle /10665/44304/9789241599283_eng.pdf?sequence=1]

[143] GISAID. Global initiative on sharing all influenza data 2020. [Available from: https://www.gisaid.org/]

CHAPTER 2

Life Cycle of Influenza Virus

Abstract: This chapter provides information on the structural organization and replication of influenza viruses in infected cells. The role of genes in the manifestation of pathogenicity and the attenuation of influenza viruses is discussed.

Keywords: Attenuation, Cold-adaptation, Drift, Genetic basis of attenuation, Genome structure, Influenza virus, Mutation, Pathogenicity, Polymerase genes, Proteins, Reassortment, Replication, RNA, Shift, Temperature sensitivity.

INTRODUCTION

The influenza virus is well adapted to reproduce in the host cell. Having a single-stranded negative-sense RNA-genome, it fully uses the resources and replicative apparatus of the infected cell for its own needs. Chapter **2** describes the structure of the influenza virus virion and its replication in a sensitive cell. The progress of molecular genetics contributes to the deepening of knowledge about the mechanisms of viral replication. This gives optimism in the success of the search for effective methods of combating influenza infection, identifying the basics of viral pathogenicity and attenuation, and brings closer to the possibilities of developing universal vaccines.

GENOME AND CAPSID STRUCTURE

Influenza A and B viruses are similar in structure (Fig. **1**) [1, 2]. Spherical virions are 80–120 nm in diameter, although filamentous formations up to several microns in length are often found, especially among fresh isolates [3].

The viral particle is covered by a lipid bilayer membrane. This membrane, like the carbohydrate components of the virus glycoproteins, is borrowed from the host cell.

Approximately 600 spines of two types are immersed with a hydrophobic end in the lipid membrane. About 500 rod-shaped spines are homologous trimers of the transmembrane glycoprotein HA (hemagglutinin), which is functionally responsible for the attachment of the virus to the host cell receptors and mediating

Irina Kiseleva and Natalie Larionova

virus entry. About 100 other spines are mushroom-shaped and consist of tetramers of the transmembrane glycoprotein NA (neuraminidase), which has enzymatic activity.

The tetramers of the M2 protein are also integrated into the envelope of the influenza A virus. This protein is a minor component of the viral particle (20-60 M2 protein molecules per virion, 1 M2 tetramer per 10-100 HA structures), but it plays a unique role as a proton pump, providing conditions for the deproteinization of the virus during its penetration into the cell [3].

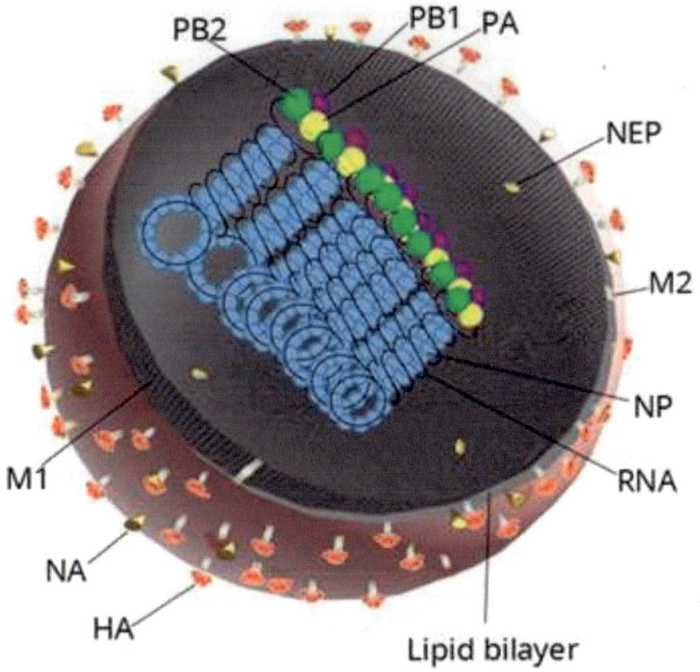

Fig. (1). Influenza A virion structure.

Influenza B virions have 4 proteins embedded in the envelope: HA, NA, NB, and BM2. The BM2 tetramer forms a proton channel, similar to the M2 protein of the influenza A virus; besides, it functions as intracellular transport of the M1-vRNP complex to the site of virion assembly [4]. The NB protein forms ion channels as well that are important for viral replication in the host cell but are not involved in viral replication *in vitro* [5].

The lipid membrane is lined with an M1 (matrix) protein from the inside. It is the largest component of the viral particle - 3000 copies of the M1 protein per virion. M1 protein is a link between the envelope and the core the virion. The cytoplasmic tails of the HA, NA, and M2 (BM2, NB - in influenza B viruses) proteins are anchored in the matrix protein and the nuclear structures vRNP (ribonucleoprotein) segments and NEP (the nuclear export protein) located inside [6].

NEP is a minor component of a viral particle that plays several biologically important roles during the life cycle of the influenza virus. NEP facilitates the export of progeny vRNA segments from the synthesis site in the nucleus to the cytoplasm and the site of viral particle assembly at the cell membrane. In this process, NEP acts as a link between the vRNP complex and the export mechanism of the cell nucleus [7, 8]. NEP is also involved in the regulation of the level of vRNA transcription and replication by viral RNA-dependent RNA polymerases and the switch of cellular ATPase to efficient budding of virions. This regulation of viral RNA transcription and replication by NEP is an important factor in the adaptation of highly pathogenic H5N1 avian influenza viruses to mammals [9].

Each of the eight discrete RNPs of influenza A and B virions includes a segment of negative-sense RNA, which is a separate gene associated with a protein complex of nucleoprotein (NP) and RNA-dependent RNA polymerase (Fig. **1**) [1, 2]. Viral RNA-dependent RNA polymerase consists of subunits of the alkaline proteins PB2, PB1 and the acidic protein PA. The polymerase complex is responsible for vRNA transcription and replication [10]. To implement its main functions, the vRNP complex can adapt to different temperature regimes of the hosts during the circulation of influenza viruses in mammals and birds [11 - 13].

Influenza C and D viruses have seven vRNP segments. The functions of hemagglutinin and neuraminidase are carried out by one polyfunctional fusion glycoprotein - hemagglutinin esterase (HEF). The homogeneous spines on the surface of virions are organized into ordered hexagonal structures that differentiate them from A and B viruses [14, 15].

The following describes the properties of influenza A and B viruses, which are the main human pathogens.

The segmented RNA of the influenza virus genome has a size of 12-15 kilobases. The electrophoresis of viral RNAs in polyacrylamide (PAAG) gel (PAGE) allows seeing the distribution of genome segments depending on their molecular weight (Fig. **2**). The migration rate of genome segments is inversely related to their molecular weight. Segments encoding viral proteins migrate in PAAG as follows: PB2, PB1, PA, HA, NP, NA, M, and NS. Different strains of the influenza virus may vary in the migration rates of their RNA segments.

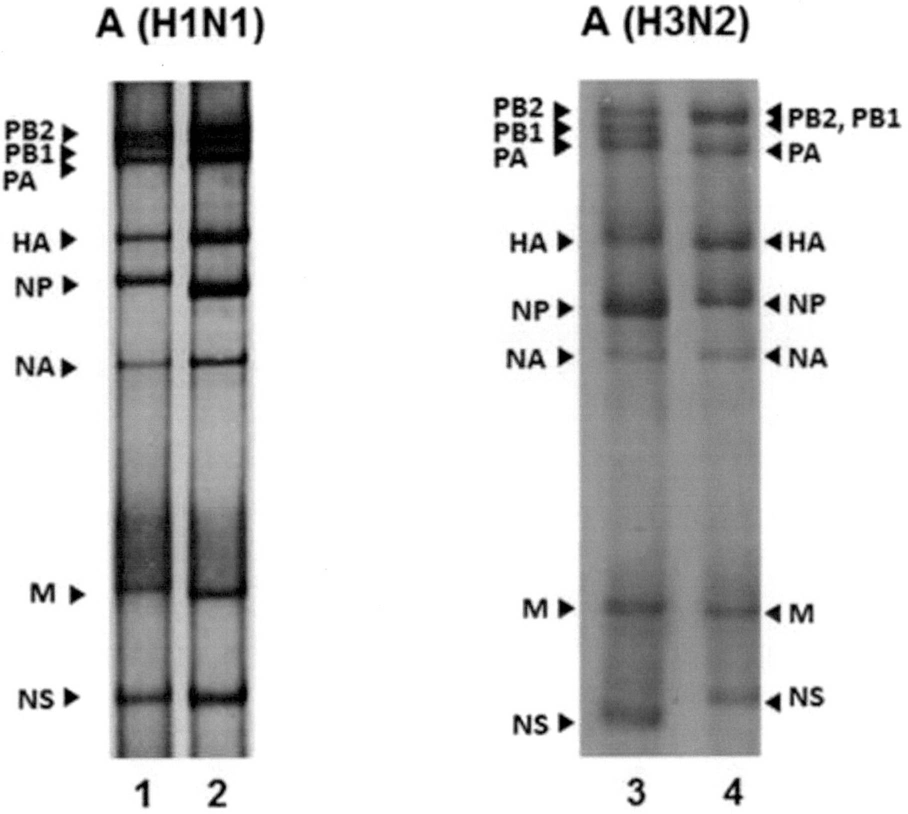

Fig. (2). The migration pattern of influenza A viruses RNA segments in PAGE *(N. Larionova, pers. comm.)*. Left panel – RNA segments of A (H1N1) influenza viruses. Right panel – RNA segments of A (H3N2) influenza viruses. Line 1 – A/England/333/80 (H1N1); line 2 – A/USSR/90/77 (H1N1); line 3 – A/Texas/1/77 (H3N2); line 4 – A/Bangkok/1/79 (H3N2).

At the 5 'and 3' ends of each vRNA segment are localized complementary sequences, due to which the RNA is twisted into a helical structure with a loop at the distal end, it is figuratively compared with a "corkscrew" or "handle of a frying pan" [3, 16]. The partially complementary base pair ends function as a promoter for the transcription and replication of vRNA by the polymerase complex. The non-coding regions include a signal for mRNA polyadenylation and a signal fragment for viral assembly. Each genomic segment encodes one or more structural and non-structural proteins. Viral proteins form the structure of the virion and ensure the replication of the virus in the host cell. Structural proteins determine the host range, tissue tropism, transmissibility, antigenic properties, and, to a large extent, the pathogenicity of the influenza virus. Non-structural proteins are present only in an infected cell, and they encode virulence factors that allow the virus to overcome the host's defense mechanisms [17, 18]. The characteristics and functions of viral proteins are described in Table **1**.

Table 1. RNA segments of influenza A and B viruses and proteins encoded by them.

RNA segment		**Protein**		
#	**length (nb[1])**	**Encoded**	**Monomer size (aar.[2])**	**Functions**
Influenza virus A, on an example of influenza A/PR8/34 virus [19]				
1	2341	PB2	759	RNA polymerase/transcriptase. mRNA recognition and binding of "cap" structures [20].
2	2341	PB1	757	RNA polymerase/transcriptase. Primer elongation, endonuclease activity [20].
		PB1-F2	87	Apoptosis regulator [21], interferon response [22] regulator, virulence factor, regulates the host sensitivity to bacterial superinfection [23].
		N40	718	Virus-induced pathogenesis in conjunction with PB1-F2, insufficiently studied [24, 25].
3	2233	PA	716	RNA polymerase/transcriptase. Endonuclease cleaves the mRNA of cells, their cap fragments become a seed in the synthesis of mRNA of the virus. Assembly of the polymerase complex [19].
		PA-X	252	Virulence factor. Disabling the mechanisms of the host cell functioning [26].
4	1778	HA	566	Hemagglutinin. Envelope glycoprotein. The main viral antigen. Cell receptor binding, fusion activator [19].
5	1565	NP	498	Nucleoprotein. Structural link with RNA, part of the transcriptase complex. Nuclear and cytoplasmic transport of vRNA [19].
6	1413	NA	454	Neuraminidase. Enzymatic envelope glycoprotein. The second most important viral antigen. Functional balance with HA [27]. Release of the virus from an infected cell during budding, cleavage from receptors, assistance in passing through the secretions of the mucous membranes of the upper respiratory tract [28].
7	1027	M1	252	Matrix protein. The main structural component of the virion interacts with the nucleus: vRNP and NEP and envelope proteins. Primary regulatory role in virus assembly and budding [29].
		M2	97	Protein is integrated into the membrane. Tetramers form a proton channel that plays an important role in undressing the virion. Participates in assembly, budding, and morphogenesis. Transport of the M1-vRNP complex to the virion assembly site [4].

(Table 1) cont.....

8	890	NS1	230	Non-structural protein of nuclear localization. Pathogenicity factor: acts as a regulator of the host cell and the virus functions: inhibits splicing, translation, and transport of cellular mRNA, the antagonist of interferon, and the activity of genes producing interferon [4].
		NEP/NS2	121	Nuclear export protein. Transport of viral RNAs from the nucleus to the cytoplasm [7], regulation of the vRNA, cRNA, and mRNA synthesis, switching of cellular ATPase to efficient budding of virions [9].
Influenza virus B, on an example of influenza B/Lee/40 virus [19]				
1	2369	PB2	770	RNA polymerase/transcriptase. mRNA recognition and binding of "cap" structures [20].
2	2368	PB1	752	RNA polymerase/transcriptase. Primer elongation, endonuclease activity [20].
3	2245[3]	PA	726	RNA polymerase/transcriptase. Endonuclease cleaves the mRNA of cells, their cap fragments become a seed in the synthesis of mRNA of the virus. Assembly of the polymerase complex [19].
4	1882	HA	584	Hemagglutinin. Envelope glycoprotein. The main viral antigen. Cell receptor binding, fusion activator [19].
5	1841	NP	560	Nucleoprotein. Structural link with RNA, part of the transcriptase complex. Nuclear and cytoplasmic transport of vRNA [19].
6	1557	NA	486	Neuraminidase. Enzymatic envelope glycoprotein. The second most important viral antigen. Functional balance with HA [27]. Release of the virus from an infected cell during budding, cleavage from receptors, assistance in passing through the secretions of the mucous membranes of the upper respiratory tract [28].
		NB	100	Protein is integrated into the membrane; ion channel [30].
7	1180[3]	M1	248	Matrix protein. The main structural component of the virion interacts with the nucleus: vRNP and NEP and envelope proteins. Primary regulatory role in virus assembly and budding [29].
		BM2	109	Protein is integrated into the membrane. Tetramers form a proton channel that plays an important role in undressing the virion. Participates in assembly, budding, and morphogenesis. Transport of the M1-vRNP complex to the virion assembly site [4].
8	1096	NS1	281	Non-structural protein of nuclear localization. Pathogenicity factor: acts as a regulator of the host cell and the virus functions: inhibits splicing, translation, and transport of cellular mRNA, the antagonist of interferon and the activity of genes producing interferon [4].
		NEP	122	Nuclear export protein [8]. Transport of viral RNAs from the nucleus to the cytoplasm [7], regulation of the vRNA, cRNA, and mRNA synthesis, switching of cellular ATPase to efficient budding of virions [8, 9].

[1]Nucleotide bases. [2]Amino acid residues. [3]Genome segment of B/Memphis/12/97 virus [19].

In the structure of vRNP, the NP protein is predominant - about 1000 copies per viral particle, and there are about 30 to 60 copies of each component of polymerase proteins. The NP in the influenza virion is organized in a trimer. On the outside, each of the monomers forms a deep pocket, adapted for binding with RNA. Binding occurs through a large number of base residues on the surface of the pocket, which interact with the phosphodiesterase backbone of RNA. The distance between two adjacent RNA binding sites in an NP trimer molecule corresponds to 70 angstroms, which is the same as one NP molecule per 24 RNA nucleotides. The crystal structure of the NP shows that it is organized into an ordered framework around which a segment of vRNA is wrapped. In addition to its structural role in vRNP organization, NP plays an important role in RNA transcription and replication. Despite the high conservatism of the NP amino acid sequence of influenza A and B viruses, there are genus genetic differences between them [31 - 33].

NP is a target to which the cytotoxic T-cell immune response of an infected (or immunized) organism is directed.

Envelope glycoproteins HA and NA are the main highly variable antigenic determinants of the virus. The rapid evolution of surface glycoproteins HA and NA enable the virus to resist the immune press of the host organism.

Hemagglutinin causes the induction of virus-neutralizing antibodies. Antibodies to neuraminidase do not neutralize the virus, but prevent the release of viral progeny from infected cells, leading to reduced replication and preventing the spread of infection. Both proteins, HA and NA, recognize the same molecule in the host cell membrane glycoproteins and glycolipids - the sialic acid (an N-substituted derivative of neuraminic acid).

HA subunits are presented on the virion membrane in the form of homotrimers and are visualized under an electron microscope as rod-shaped spines [34]. Initially, HA is synthesized as a single precursor polypeptide (HA0) and combined into a homotrimer. Viral HA contains an N-terminal signal sequence and a C-terminal membrane anchor. The native HA0 trimer is cleaved by cellular proteases into the HA1 and HA2 subunits, held together by a disulfide bond [34]. The N-terminus of HA2, which contains the fusion peptide, is slightly displaced from the previously adjacent C-terminus of HA1. The HA1 trimer forms a globular head, and three HA2 subunits with a part of HA1 form the stem [35]. In contrast to the high variability of the amino acid sequence in the head of HA, the stem is characterized by a conserved structure in various strains and serotypes of the virus.

After synthesis and proteolytic cleavage, the HA molecule undergoes glycosylation and acylation. Glycosylation occurs at 5–7 sites. The amount and positions of carbohydrate chains attached to HA vary in different strains during the evolution of the pathogen [36, 37]. Glycosylation of the globular head of HA modulates the antigenicity, fusion activity, virulence, receptor-binding specificity, and immune evasion of the influenza virus [38]. During acylation, palmitic acid is attached to conserved cysteine residues located in the transmembrane C-terminal region of HA, thus forming hydrophobic regions involved in the anchoring of HA in the envelope and concentration of HA at budding sites during assembly [39].

Proteolytic cleavage and reorganization of the initial HA0 are required for the virus to conformationally transition to an active state for fusion with the endosome – a factor of viral infectivity [40]. The fusion of the viral and endosomal membranes is initiated when the endosome is acidified to pH 5-6.

In the recess of the HA head, is located a receptor-binding site, which consists mainly of conservative amino acids. Due to it, the attachment of the virus to the cell receptors and the subsequent fusion of the viral membrane and the endosome of the cell is ensured. The receptor-binding site is surrounded by variable antigenic determinants. Mutations in antigenic determinants allow the virus to evade the host's humoral immune response [41]. In influenza viruses of serotype H3, they are designated as A, B, C, D, and E, and in viruses of serotype H1, as Sa, Sb, Ca1, Ca2, and Cb [20]. In influenza B viruses, one large antigenic site is formed by spatially overlapping epitopes from the 120th, 150th, 160th loops, and 190th helix. Despite the low, approximately 20%, homology in the HA sequences of influenza A and B viruses, they are characterized by structural similarity [42].

The functions of NA, which has enzymatic activity, are associated with the cleavage of neuraminic acid from the adjacent sugar residue in glycoprotein complexes, through which viral NA interacts with cellular structures. At the stage of the virus enter, NA destroys the receptor to which HA has attached, facilitating the fusion of the viral and cell membranes [28]. At the later stages of infection, during budding, NA prevents the self-aggregation of viral particles, which ensures the free release of newly synthesized virions from the infected cell and their spread to neighboring cells [43]. Besides, upon infection, NA promotes the progression of the viral particle to cellular receptors, liquefying neuraminic acid-containing mucin secretions of the respiratory tract tissues [27].

The NA monomer consists of a cytoplasmic and transmembrane region, the head, and a stem connecting the head to the transmembrane region. On the globular head are the enzymatic and antigenic sites of the molecule. In virions, NA forms a homotetramer from subunits linked in pairs by disulfide bonds. Active sites are located in a deep pocket on the outer surface [44].

Carbohydrate chains attach to aspartic residues in different regions of the NA head, which vary among influenza viruses. At the same time, the Asn146 glycosylation site is conserved in all influenza viruses; it seems to play a regulatory function since its absence in A/WSN/33 (H1N1) leads to neurovirulence [44].

Numerous studies in recent years have shown that effective viral replication requires an optimal relationship between the receptor-binding and receptor-destroying activities of surface glycoproteins. The balance existing in the virus between the functional antagonists of HA and NA can be disturbed by various events: reassortment, transmission of the virus to a new host, inhibition of neuraminidase as a result of antiviral therapy. A decrease in viral replicative compliance is usually overcome by restoring functional balance through compensatory mutations in HA, NA, or both [27].

To date, it is known that the genome of the influenza A virus, in addition to 9 structural proteins, is capable of encoding at least 8 non-structural proteins (Table **1**). Segments 2 and 3, the primary protein products of which are polymerase proteins PB1 and PA, respectively, additionally encode proteins PB1-F2, PB1-N40, and PA-X. The influenza A virus PB1-F2 and PB1-N40 proteins and the influenza B virus NB protein are encoded through overlapping reading frames (PB1-F2 - with a +1 shift, N40 is translated through the same open reading frame as PB1, but initiated from the second AUG codon, protein NB - from an alternative reading frame -1). The unspliced 7th and 8th mRNA segments encode the M1 and NS1 proteins, respectively, while the M2 proteins of the influenza A virus and NEP of both viruses A and B are encoded by the 7 and 8 segments as a result of differentiated mRNA splicing [25]. Other spliced mRNAs encode the recently identified M42 protein and the undetected yet M proteins, which are probably involved in the early stages of infection but are insignificant for the growth of influenza A virus in cell culture [45]. The BM2 protein of the influenza B virus is encoded by segment 7 through a progressive stop-start mechanism [46].

Not every strain expresses all non-structural viral proteins identified in recent years. The absence of these proteins does not affect the viability of the virus, but can adversely affect its replication. The functions of recently discovered non-structural proteins have not been adequately studied yet. The genome of the influenza B virus encodes 11 proteins – 10 structural and one non-structural protein [1]. Due to the activity of the NS1 and PB1-F2 proteins, the influenza virus has mechanisms that allow it to resist cellular factors of antiviral defense. In particular, NS1 is the main antagonist of interferon; this protein also counteracts some other antiviral factors of the cell; it also takes part in the regulation of cell death (apoptosis).

The PB1-F2 protein is present in most but not all influenza A viruses. It is a short protein, its length varies from 55 to 90 amino acids in different viruses, and in the vast majority of H1N1pdm09 virus isolates, the truncated PB-F2 protein consists of only 11 amino acids. The PB1-F2 protein destabilizes mitochondrial membranes due to the formation of pores in them, which leads to leakage of macromolecules and ends with apoptosis [21, 22, 47, 48]. PB1-F2 also affects the activity of the polymerase complex by interacting with the PB1 protein [49]. All the results of the activity of the PB1-F2 protein are in the increase of virus pathogenicity.

REPLICATION IN HOST CELL

To implement the functions of reproduction, due to its own small coding capabilities, the influenza virus uses different strategies: in addition to the synthesis of several polypeptides from one segment of the genomic RNA, the virus usurps and modifies the replicative apparatus of the host cell; besides, several viral proteins have polyfunctional properties [3, 50, 51].

The Entry into the Cell

The initial stage of the replication cycle is the adsorption of the virus on the surface of the host cell [1, 2]. The influenza virus recognizes N-acetylneuraminic (sialic) acid in the cell membrane as a receptor.

Sialic acids are nine carbon monosaccharides found at the end of many glycoconjugates. They are widely distributed in different types of animal cells. The second carbon in the terminal sialic acid can bind to either the third or sixth carbon of galactose, forming Siaα-2,3Gal (α-2,3) or Siaα-2,6Gal (α-2,6) bonds, respectively. Differences in the mode of binding result in unique steric configurations of the terminal sialic acid. These sialic acid groups are recognized by influenza virus HA receptor binding sites that have specificity for the α-2,3 or α-2.6 linkages. Human influenza A viruses recognize as cellular receptors glycoproteins containing sialic acids bound to galactose through the α-2,6 bond [52 - 55]. Such sialyl-containing receptors predominate on epithelial cells of the human upper respiratory tract [28, 56, 57]. Avian and equine influenza viruses preferentially bind to α-2,3 receptors [52 - 55], which, in particular, predominate on duck intestinal epithelial cells [58, 59]. Swine influenza viruses show an affinity for both types of receptors [60]. These differences in the receptor specificity of HA are a crucial determinant of the host range of influenza viruses [20, 61]. As for influenza B viruses, which are practically endemic for humans, the viruses of the Yamagata-line have a predominant affinity for receptors with α-2,6-bond, while the viruses of the Victoria-line can bind to both types of receptors - with α-2.6 and α-2,3 bond; besides, the third type of receptor is specific for their representatives – sulfated glycoproteins [62].

Endocytosis and Intracellular Migration of Virus-Containing Endosomes

Upon virus entry, HA is activated by low pH and undergoes large structural rearrangements from a prefusion state to a postfusion state. When the influenza virus attaches to the plasma membrane, it induces the formation of claritin-coated depressions that are converted into endosomes [63].

Intracellular transport of endosomes containing the virus occurs in three stages. The first stage is the slow movement of endosomes with the virus to the periphery of the cell. In the second stage, endosomes with the virus quickly move to the cell nucleus. During the third stage, the virus moves back and forth along the same path in the perinuclear region. It has been suggested that as a result of this movement, an endocytic "corridor" is formed, which organizes multidirectional virus-specific motor actions. The final stage of this process is the fusion of the viral envelope with the endosome membrane.

The fusion takes place in the perinuclear area; here, due to the action of the proton pump, acidification of the endosome occurs. At a low pH, HA undergoes irreversible conformational changes, which result in the release of a "fusion peptide" at the N-terminus of the HA2 subunit. HA acquires the ability to integrate into the endosomal membrane, which leads to its fusion with the viral membrane, and the formation of a fusion pore that enables the viral ribonucleoprotein particles to be released into the cytoplasm for viral replication [40].

As HA plays a key role in the entire entry pathway, inhibitors of virus entry are potentially effective antiviral drugs of influenza viruses [40, 64]. Such a broad-spectrum antiviral compound is an inhibitor of the fusion process of the influenza virus Arbidol (ethyl-6-bromo-4 - [(dimethylamino) methyl] -5-hydroxy-1-me-hyl-2 - [(phenylthio) methyl] -indole-3-carboxylate hydrochloride monohydrate). Arbidol is approved for use in healthcare in Russia [65]. The deep mechanism of action of Arbidol continues to be studied [66].

The pH in the endosome depends on the host species, and the pH stability of HA is one of the determinants of viral tropism [67].

Through the ion channel formed by the viral protein M2 (BM2), an influx of protons into the viral particle occurs, and it also acidifies. This leads to the dissociation of viral vRNP complexes from the matrix protein into the cytoplasm and their transport to the cell nucleus [68]. This process involves the cellular nuclear import factors importin-α and importin-β [69]. The M1 protein is independently imported into the nucleus [68].

Transcription and Replication of the Viral Genome

Unlike most other RNA viruses in orthomyxoviruses, all virus-specific RNAs are synthesized in the nucleus of the infected cell. The process is catalyzed by a virus polymerase complex of three subunits PB2, PB1, and PA. Viral RNA serves as a template for the synthesis of cRNA (complementary) and mRNA (template).

Replication of viral RNA is carried out through a two-step process: it begins with the synthesis of complementary cRNA, which is a (+) sense copy of vRNA. Further, cRNA is copied many times and serves as a template for the synthesis of a large number of daughter vRNAs [70, 71].

vRNA transcription is initiated by binding the PB2 protein to the cap-structure of cellular mRNA. Cap structures act as a primer for mRNA synthesis. Virus mRNA is synthesized with the participation of the PB1 protein. Transcription proceeds until the work of the polymerase complex freeze at the polyadenylated signaling site at the end of the viral RNA [72].

Translation

The synthesized viral mRNAs are exported from the nucleus to the cytoplasm, where a translation of virus-specific proteins occurs, while the cellular translation mechanisms are readjusted for the needs of the infectious agent [73, 74]. The HA protein monomer is synthesized as a precursor polypeptide - HA0. The synthesized proteins of the polymerase complex – NP, PB2, PB1, and PA – are imported into the nucleus using nuclear localization signals [70]. In the nucleus, they catalyze vRNA transcription and replication. Proteins M1, NEP, and NS1 are also directed to the nucleus to fulfill their role in the export from the nucleus of vRNA (M1 and NEP) or the processing and export of cellular and viral mRNA (NS1) [70, 75].

Export of vRNA

To export newly synthesized vRNA sets, viral proteins NEP and M1 work in interaction with cell proteins that form a "bridge" between vRNP and NEP, and create an association between the M1 protein and vRNA [70].

Post-Translation Modifications

After synthesis in the cytoplasm, HA, NA, and M2 proteins enter the endoplasmic reticulum, where HA0 is assembled into a trimer, proteins HA and NA undergo glycosylation, HA and M2 - S-acylation, M1, NS1, NP, PB1, and NEP conjugate with a small ubiquitin-like modifier activating enzyme (SUMOylation). Phosphorylation of M1 and NP proteins, which can affect nuclear import and export of vRNA, and phosphorylation of NS1 and PB1-F2 proteins, which affects virulence, is carried out. All these processes are catalyzed by cellular kinases [70]. Through the Golgi transmembrane network, the HA trimer is exported to the cell surface, where cellular proteases cleave the HA protein into disulfide-linked HA1 and HA2 subunits. The HA protein of highly virulent avian influenza viruses of H5 and H7 serotypes possesses several basic amino acids in the HA cleavage site, which differs it from human viruses and non-pathogenic avian viruses, in which the HA cleavage site has one main amino acid, arginine [76, 77].

The Transport of Viral Proteins to the Plasma Membrane

The transport of viral proteins through the plasma membrane is performed by cell factors.

Packaging and Budding

In the plasma membrane, HA and NA bind to lipid rafts (membrane regions rich in sphingolipids and cholesterol), which become the site of influenza virus budding [78]. The M1 protein is involved in the assembly of the virus – it interacts with the lipid membrane, vRNA, and NEP [6, 29, 50, 69, 79]. The cytoplasmic tail of the M2 protein may also be involved in the insertion of vRNA into a viral particle [80].

When assembled, the daughter virions always contain a complete set of eight RNPs and no duplicates, which indicates the existence of a segment-specific signal for packaging. However, the mechanisms of assembly regulation have not yet been sufficiently studied [16, 81 - 84].

The viral NA protein cleaves sialic acids from the host cell membrane and the HA virion glycoproteins. Daughter viral particles are released and can infect neighboring cells [70].

MOLECULAR MECHANISMS OF ATTENUATION

Since the days of Louis Pasteur, it has been known that passages of the pathogen through another host species leads to attenuation. However, molecular mechanisms became known only with the development of molecular biology, and some are not known even today.

Basic About Cold-Adapted LAIVs

Influenza virus affects up to 20% of the global population. The fatality is extremely high among certain groups of the population - children, elderly people, pregnant women, and people with a weak immune system (high-risk groups). Serious threats to humankind are flu pandemics, which occur 3-4 times per century and are characterized by a very rapid spread covering whole continents or the whole of the globe with the incidence of disease percent far exceeding normal and high fatality in all population groups.

As of today, influenza vaccines are one of the most effective ways to protect people from contracting an illness during influenza epidemics and pandemics. According to WHO, there is evidence that live, attenuated cold-adapted intranasal reassortant influenza vaccines (LAIVs) might be more effective than inactivated vaccines, especially in the pandemic situation [85]. LAIVs are currently in the market in the Russian Federation and the United States and appear to be safe and efficacious (for more details, see Chapter **4**). Safety and effectiveness of live needle-free live influenza vaccines were confirmed by hundreds of clinical trials. They may also provide long-lasting and broader immune (humoral and cellular) responses than inactivated needle-injected vaccines [86].

The current LAIV is a trivalent or quadrivalent vaccine composed of influenza A virus subtypes H1N1 and H3N2 and influenza B (Victoria-line or/and Jamagata-line) virus. The Russian LAIV Ultravac® is based on the master donor viruses (MDVs) A/Leningrad/134/17/57 (H2N2) (L17) and B/USSR/60/69 (B60) that have been rendered temperature-sensitive (*ts*) and cold-adapted (*ca*) by successive passages in eggs at 32°C and 25°C, respectively [87]. These MDVs are used in reassortment with current wild type (WT) influenza A and B viruses, respectively, to generate attenuated (*att*) vaccine viruses containing the proper surface glycoproteins. Ultravac® is licensed in Russia for use in healthy individuals aged over 3 years. FluMist®, a LAIV that is available since 2003 in the USA and licensed for use in healthy individuals aged 2-49 years, is based on the cold-adapted MDVs A/Ann Arbor/6/60*ca* (H2N2) (A/AA) and B/Ann Arbor/1/66*ca* (B/AA), which were developed independently from the Russian MDVs but using the same strategy [88] (for more details see Chapter **4**).

The *ca/ts/att* phenotype of LAIV is inherited from the six mutated genes encoding the internal proteins of the MDV. In the next sections, we will discuss what role each mutant gene plays in the attenuation of cold-adapted viruses, MDVs, and LAIVs.

What Role Do Mutations/Mutated Genes Play in Attenuation?

The acquisition of genes from the MDV into the background of a WT virus has been shown to reduce the virulence of the WT virus. This is the classical approach towards the development of live attenuated influenza vaccines. Single gene reassortant (SGR) and multiple gene reassortant studies provide an opportunity to analyze the effect of variant gene segments from *ca* influenza viral strains into the background of WT influenza viruses [89 - 91].

More precise experiments with single-gene reassortants (SGRs) between genetically related viruses, pointed towards a leading role for mutated genes in conferring the *ts/ca* phenotype. To more easily identify the genes that are responsible for the temperature sensitivity of Ultravac® A- and B- MDVs, reassortants containing one, two, three, or four genes derived from the L17 MDV with the remaining genes derived from its WT progenitor, A/Leningrad/134/57 influenza virus were generated and studied. The systematic exchange of single mutant genes into the background of WT genes by classical reassortment has allowed for the determination of the roles of each mutant gene of Ultravac® A- and B-MDVs towards conferring *ts*, *ca*, and/or attenuation [92 - 94].

An alternative approach - reverse genetics - allows engineering unique amino acids into a plasmid-derived recombinant of FluMist® A- [95] and B-MDV [96] and Ultravac® A-MDV [97] to find out what particular amino acid changes will revert *ts/att* phenotypes.

Eight coding mutations identified in the genes coding for the internal proteins are present in PB2 (1), PB1 (2), PA (2), M1 (2), and NS2 (1) genes of L17 Ultravac® MDV [97]. Compared to L17 MDV, the A-MDV used in FluMist®, A/AA contains 10 mutations in the genes coding for the internal proteins, eight of which are in RNP encoding genes (PB2, PB1, PA, and NP) [95]. No substitutions on the amino acid level in the M1 and NS2 gene segments in A/AA were found (Table **2**).

Eight coding mutations identified in the genes coding for the internal proteins are present in PB2 (3), PA (2), M1 (1), BM2 (1), and NS1 (1) genes of B60 MDV. Of the 8 coding mutations (2 non-unique and 6 unique mutations), five are in genes coding for the RNP complex (PB2 and PA) [93]. Compared to B60 MDV, the B/AA MDV used in FluMist® contains 9 mutations in the genes coding for the internal proteins (1 non-unique and 8 unique mutations), seven of which are in RNP encoding genes (PB2, PA, and NP) [96]. No substitutions on the amino acid level in the PB1 and NP genes in B60 MDV were found (Table **2**).

Table 2. Amino acid changes in the internal proteins of Ultravac® MDVs compared to FluMist® MDVs and their role in the temperature sensitivity of MDV.

		Number of AA changes	*ts* phenotype	Number of AA changes	*ts* phenotype
Gene segment		**L17 MDV for Ultravac [97, 98]**		**A/AA MDV for FluMist for [95]**	
PB2		1	*ts*	1	*ts*
PB1		2	*ts*	4	*ts*
PA		2	*non-ts*	2	*non-ts*
NP		No changes	*non-ts*	1	*ts*
M	**M1**	2	*non-ts*	No changes	*non-ts*
	M2	No changes		1	
NS	**NS1**	No changes	*non-ts*	1	*non-ts*
	NS2	1		No changes	
RNP-complex		5 unique mutations		8 unique mutations	
Gene segment		**B60 MDV for Ultravac [93][1]**		**B/AA MDV for FluMist [96][1]**	
PB2		2	*ts*	1[1]	No role[2]
		1[1]			
PB1		No changes	*non-ts*	No changes	*non-ts*
PA		2	*ts*	2	*ts*
NP		No changes	*non-ts*	4	*ts*
M	**M1**	1	*non-ts*	2	Possibly *ts*
	BM2	1[1]		No changes	
NS	**NS1**	1	*non-ts*	No changes	*non-ts*
	NS2	No changes		No changes	
RNP-complex		4 unique mutations and 1 non-unique mutation		6 unique mutations and 1 non-unique mutation	

[1]Mutation is non-unique. [2]Mutation/mutated gene plays no essential role in *the ts/att* phenotype.

Identification of key residues involved in *ca* and *ts* phenotype. A little is published about the role of mutations in the cold-adapted phenotype of *ca* viruses. Publications concerning the molecular basis of attenuation of cold-adapted influenza vaccines put the main focus on *ts* phenotype and attenuation of tested cold-adapted viruses [95 - 97, 99].

The early attempts to identify the genes of the A/AA and B/AA cold-adapted MDVs that impart the *ca* phenotype to reassortant vaccine viruses utilized classical genetic reassortment between cold-adapted MDV and genetically divergent epidemic viruses. PB2 and PA gene segments inherited from A/AA [90, 91, 100] and PA gene segment derived from B/AA [89] each expressed determinants of cold-adaptation.

Publication concerning Russian cold-adapted LAIV described a few amino acid substitutions which may play an essential role in the manifestation of the *ca* phenotype [92]. It was demonstrated that PB2-SGR, and PA-SGR, based on L17 MDV, each manifested the *ca* phenotype. Mutations in other genes coding internal proteins did not affect the *ca* phenotype of corresponding SGRs.

As for the *ts* phenotype, mutations in the polymerase proteins were shown to play a crucial role in the *ts* phenotype. By using the SGR technique, it was demonstrated that PB2 is the critical mutant gene in conferring temperature sensitivity in the L17 MDV. However, the other polymerase genes (PB1 and PA) are also essential in maintaining *ts* in the background of non-homologous HA and NA gene constellations [92, 94]. Further studies involving reverse genetics and site-directed mutagenesis confirmed the role of individual mutations in the *ts* phenotype of L17 MDV [97].

Polymerase PB2 and PA mutated gene segments of B60 MDV independently controlled expression of the *ts* phenotype of B60 MDV-based reassortant viruses. The other genes coding for internal proteins seems to play no role in this respect [93].

SGRs allow us to analyze the effect of the inclusion of single mutated gene segments from cold-adapted influenza strains into the background of the WT influenza virus and to identify the viral genes that determine the phenotype of interest. However, a number of key internal genes of cold-adapted viruses carry more than one amino acid change. To determine the specific residues responsible for the expression of the *ts* phenotype by the L17 MDV, reverse-genetics viruses including one or more amino acid substitutions from the *ca* L17 virus into its WT progenitor, A/Leningrad/134/57 background were generated [97]. Three mutations in PB2 and PB1 genes were of particular interest since their SGRs showed the most significant alteration of virus replication at restrictive temperatures [92]. The separate introduction of the PB1 (K265N) or PB1 (V591I) amino acid substitutions into the WT A/Leningrad/134/57 virus increased the temperature sensitivity of resulting viruses differently. The mutation at position 591 in the PB1 gene seems to contribute to a lesser extend as compared to the mutation at position 265 [97].

A plasmid-based reverse genetics system was used to create and characterize the genetic basis of the *ts* phenotype of a number of recombinant viruses between A/AA MDV and its WT *non-ts* A/Ann Arbor/6/60 progenitor. Site-directed mutagenesis analysis of the FluMist® A/AA MDV demonstrated that PB2, PB1, and NP genes derived from A-MDV each expressed determinants of temperature sensitivity. The following coding mutations contributed to the *ts* phenotype: PB1 (K391E, E581G, A661T), PB2 (N265S), and NP (D34G) [95]. Compared to L17 MDV, it is interesting to note that the same polymerase genes were responsible for attenuation.

Eight unique amino acids in three internal genes are known for B/AA MDV: two in PA (V431M, Y497H), two in M1 (Q159H, M183V), and four in the NP gene segment (A55T, V114A, P410H, A509T). To define gene segments responsible for *ts/att* phenotypes of B/AA MDV for FluMist® reverse genetics system was used. A detailed mutational analysis revealed that a combination of two residues in NP (V114A and P410H) and one in PA (V431M) controlled the expression of *ts*, whereas these same changes in combination with two additional residues in M1 (Q159H, M183V) controlled the *att* phenotype in ferrets.

The fact that cold-adapted Ultravac® and FluMist® MDVs have the strongest *ts* determinants in the polymerase proteins suggests a critical role of those mutations in the alteration of viral polymerase activity at restrictive elevated temperatures.

What role do *ts* and *ca* phenotypes play in the attenuation of MDV? Since *ca* reassortant LAIV inherits their HA and NA genes from non-attenuated epidemic viruses, it is clear that mutations in internal genes of MDV are responsible for the attenuation.

Mutations in polymerase genes are essential for attenuation of the L17 MDV. The V478 substitution in the PB2 gene as well as K265N and V591I amino acid changes in the PB1 gene independently contribute to the attenuation of the L17 MDV for ferrets. Importantly, mutations in both these genes are responsible for the manifestation of the *ts* phenotype. On the contrary, although mutations in the PA gene (L28P and V341L) were responsible for the *ca* phenotype, they did not lead to the attenuation of the corresponding SGR for ferrets [92, 94] (Table **3**). These data demonstrated a close relationship between attenuation and *ts* rather than the *ca* phenotype of cold-adapted viruses. The PB2 gene of A/AA [99], and even some avian influenza viruses [101] have been shown to be related to the attenuation of reassortant viruses.

Table 3. Temperature-sensitive, cold-adapted and attenuated phenotypes of single-gene reassortants of L17 MDV with its WT progenitor, A/Leningrad/134/57.

Virus / SGR	Amino acid changes	Phenotype		
		ts	*ca*	*att*
LWT[1]	No changes	*non-ts* [92, 97]	*non-ca* [92]	*non-att* [92]
L17 MDV[2]	Eight changes	*ts* [92, 97]	*ca* [92]	*att* [92]
PB2-SGR	V478L	*ts* [92, 97]	*ca* [92]	*att* [92]
PB1-SGR	K265N; V591I	*ts* [92, 97]	*non-ca* [92]	*att* [92]
PA-SGR	L28P; V341L	*non-ts* [92, 97]	*ca* [92]	*non-att* [92]
NP-SGR	No changes	*non-ts* [97]	ND[3]	ND[3]
M1-SGR	I15V; F144L	*non-ts* [92, 97]	*non-ca* [92]	ND[3]
NS2-SGR	M100I	*non-ts* [92, 97]	*non-ca* [92]	ND

[1]A/Leningrad/134/57 is a WT progenitor of [2]L17 MDV. [3]Not determined.

Other studies have shown that the PA gene is responsible for the *ts* phenotype conferring attenuation to reassortant viruses based on B/AA MDV [89]. It was demonstrated that a combination of two sites in NP, one in PA, and two additional residues in M1 controlled the *att* phenotype of B/AA MDV [96]. PA gene as well as the PB2 gene of B60 MDV are independently responsible for the *ts* phenotype and are likely to be the determinants of attenuation [93].

Not all mutations were found to contribute equally to temperature sensitivity. This fact suggests that the number and/or combination of mutations determine the level of attenuation. The combination of polymerase genes seems to reinforce the manifestation of *ts* or *ca* phenotypes and may explain the observed genetic stability of Ultravac® and FluMist® cold-adapted reassortant live attenuated vaccines during their replication in humans [92, 94, 95]. This might be direct evidence that the attenuated phenotype of cold-adapted influenza viruses is under the control of multiple gene segment function.

Thus, the *ts* phenotype of the cold-adapted virus is a critical determinant for attenuation. A direct relation between attenuation and the *ts* phenotype of cold-adapted viruses has been demonstrated. In contrast, there seems to be no correlation between mutations that are responsible for the manifestation of the cold-adapted phenotype and attenuation.

A correlation between *ts* and *att* phenotype makes sense, as replication of *ts* viruses is limited to the upper respiratory tract where lower temperatures predominate. In contrast, *non-ts* viruses can grow in the warmer climate of the lower respiratory tract, where they potentially could cause viral pneumonia or some other complications.

What role do mutant genes play in the attenuation of genetically distinct WT virus? In the above paragraphs, the role of "cold" mutated genes of MDVs introduced in the genome of its WT progenitor was discussed. For a more convincing understanding of the mechanisms of attenuation, it is not enough to study only the effect of individual genes of the WT progenitor, which do not contain attenuating mutations, on the MDV phenotype. The constellation of genes in the genome of the reassortant virus may play an important role, therefore it was important to identify the "cold" genes of MDV that are responsible for the attenuation of genetically distinct WT viruses.

Kiseleva *et al.* [102] analyzed phenotypical properties of reassortants of A or B WT influenza viruses with L17 or B60 MDVs, correspondingly. The analysis was performed by comparison of their capability to grow at different temperatures in chicken eggs or/and MDCK cells. 178 reassortants of L17 with current *non-ts* influenza A WT viruses and 33 reassortants of B60 with current *non-ts* influenza B WT viruses were produced in eggs or MDCK cells and evaluated in order to identify the genes responsible for their *ts* phenotype. Reassortants that inherited at least two polymerase genes (PB2 and PA or PB1) from L17 MDV regularly demonstrated the *ts* phenotype. The polymerase PA and PB2 gene segments of B60 independently controlled the manifestation of the *ts* phenotype of B60-based reassortants. The other genes coding for internal proteins played no important role in the manifestation of the *ts* phenotype of both A17- and B60-based reassortants. These results suggested that mutated polymerase genes play a critical role in the attenuation of both A and B influenza reassortant viruses.

If the results of replacement of single mutant "cold" genes in the genome of MDV with WT genes are summarized, it could be seen those polymerase genes play the leading role in attenuation [92, 93, 97, 102] (Fig. **3**, left panel).

Not surprisingly, only the genes coding for the ribonucleoprotein complex are responsible for the manifestation of the temperature sensitivity of both Ultravac® B-MDV (PA and PB2 genes), Ultravac® A-MDV (PB2 and PB1 genes), FluMist® B-MDV (PA and NP genes), and FluMist® A-MDV (PB2, PB1 and NP genes).

No identical mutations in the genome of influenza A and B master donor viruses used in the Russian LAIV and FluMist® were found, which means that the molecular basis for the expression of the temperature-sensitive phenotype of cold-adapted master donor viruses for LAIV is close and lies within the genes that control the replication of influenza virus.

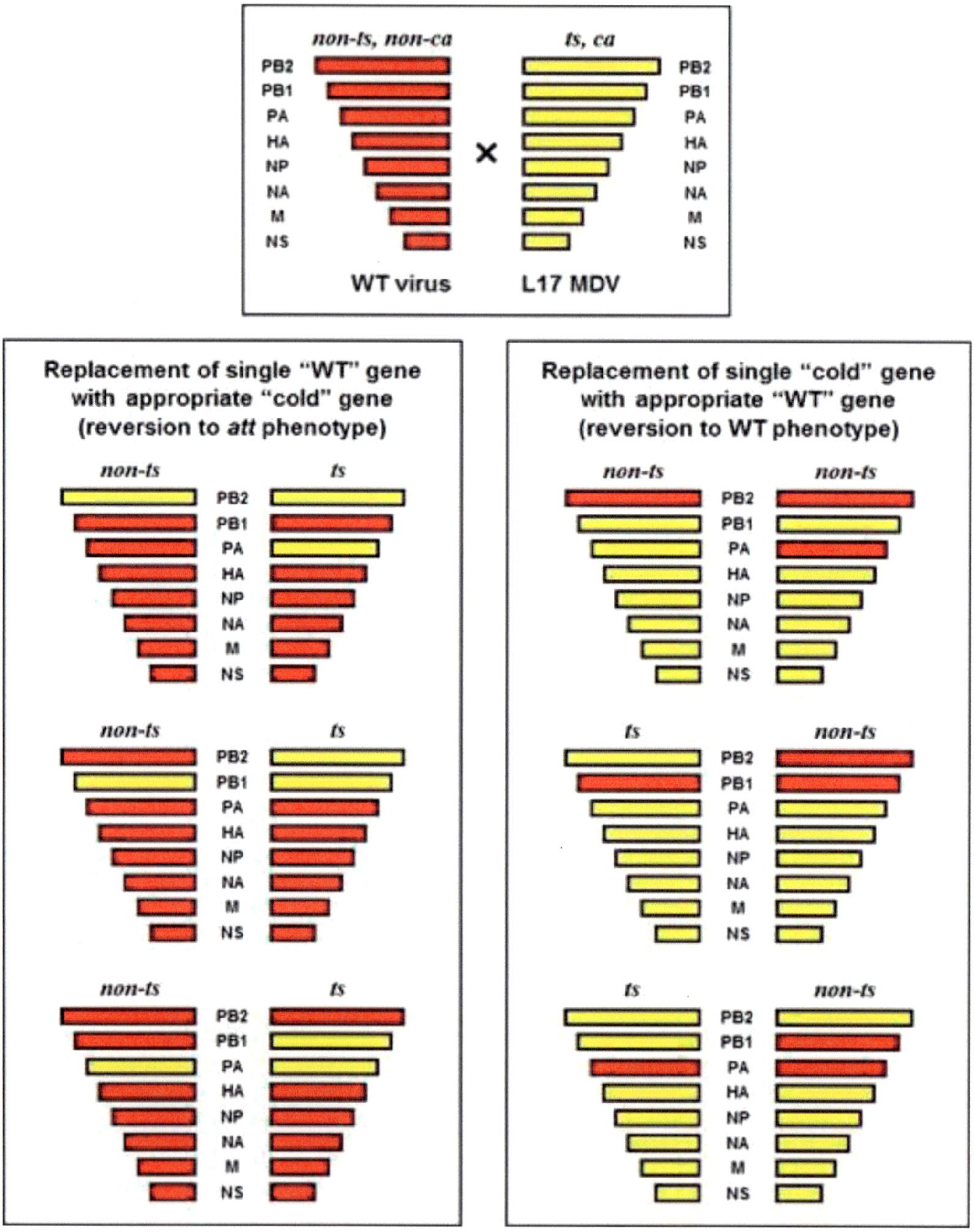

Fig. (3). The results of the replacement of single genes in the genome of MDV or WT virus with corresponding WT or mutant "cold" gene. Top panel - reassortment of recent WT influenza A virus with L17 MDV. Left panel - introducing "cold" mutated genes of L17 MDV into the genome of recent WT influenza A viruses. Right panel - introducing WT genes of recent WT influenza A viruses to the genome of L17 MDV.

This speculation allows announcing not only the role of unique mutations in the attenuation of cold-adapted influenza viruses but the role of specific genes. This suggestion is also confirmed by the results of transference of the amino acid genetic signature of the A-MDV *ts* phenotype into the divergent WT virus [97, 99, 103].

Transferring a genetic signature for the *ts/att* phenotype to the divergent WT virus. In the above paragraph, the role of "cold" mutated genes of MDVs introduced in the genome of recent WT influenza viruses was discussed. To understand how modification at the genetic level would result in the *att* phenotype of the WT virus, the involvement of key *ts* mutations in polymerase genes known for A-MDVs in attenuation of a divergent influenza virus was investigated. The genetic signature of the A-MDVs *ts* phenotype was transferred into the WT A/Puerto Rico/8/34 (H1N1) (PR8) virus [95, 97, 99, 103].

The *ts* phenotype of FluMist® cold-adapted A-MDV, A/AA, was mapped to the following five amino acids in viral polymerase genes: PB2 (N265S), PB1 (K391E, E581G, and A661T), and NP (D34G) [95]. The four of these *ts* loci were transferred into a divergent not attenuated PR8 virus [99]. The results have shown that transferring the amino acid genetic signature of the A/AA MDV *ts* phenotype into the PR8 strain imparted temperature sensitivity *in vitro* and attenuation in ferrets.

Analogous studies have been done for another cold-adapted A-MDV, L17 [97]. To determine whether the *ts* mutations of L17 Ultravac® MDV can impart the *ts* phenotype to a divergent virus, as it has been demonstrated for the A/AA MDV [99], PR8 variant viruses with different combinations of *ts* mutations from L17 MDV were created by reverse genetics. Introducing amino acid changes within polymerase genes of the PR8 strain (V591I in PB2, K265N, and V591I in PB1) led to the loss of the WT phenotype of the PR8 virus. Unlike polymerase genes, the mutation M100I in the NS2 gene did not affect the *ts* phenotype of the PR8 virus, suggesting that this mutation may only play an accessory role.

He *et al.* [103] reproduced experiments described in [97, 99] and confirmed the involvement of key mutations in the PB1 and/or PB2 genes in the attenuation of influenza A virus *in vitro* and *in vivo*. The authors have demonstrated that a few residues in PB1 and PB2 are critical for the phenotypes of cold-adapted, temperature-sensitive, and attenuated master donor viruses.

Both Ultravac® and FluMist® MDVs are based on Asia pandemic viruses, A/Singapore/1/57 (H1N1)-like strains. Romanova *et al.* [104] described a new A-MDV also based on A/Singapore/1/57 (H1N1) influenza virus. This A-MDV contains eight coding mutations identified in the genes coding for the internal proteins PB2 (I185T, R340I), PB1 (L419I), PA (N228I), M1 (R101K, Q158H). All these amino acid substitutions are different from those of Ultravac® and FluMist® MDVs. However, we can speculate with a certain degree of confidence that transference of these mutations into the genome of divergent non-attenuated influenza virus may also lead to its attenuation.

An Alternative Method of Attenuation of LAIV

The development of reverse genetics techniques allows fantastic opportunities to generate unique modifications of influenza viruses, including a novel type of live attenuated influenza vaccines. For instance, the nonstructural protein NS1 of the influenza virus counteracts the interferon-mediated immune response. By deleting the open reading frame of the NS1 gene segment, a novel type of intranasal attenuated influenza vaccine was generated. Modification of the only NS gene of the WT influenza virus led to its lasting attenuation. The use of NS1-modified WT influenza viruses as live influenza virus vaccines provide an alternative to the LAIV strategy. WT influenza viruses lacking the viral interferon antagonist NS1 induces strain-specific and cross-neutralizing antibodies despite the highly attenuated replication-deficient phenotype [105 - 107].

While the attenuated phenotype of NS1-modified viruses was demonstrated in different animal models (mice, pigs, horsed, macaques, birds) [105] and was well tolerated in humans [107], the question of their potential to revert to a virulent phenotype is still open. The theoretical possibility of reassortment of attenuated NS1-modified virus with currently circulating viruses is of serious concern if the WT virus will contribute gene segments (especially NS gene segment) to the NS1-modified live vaccine virus. In contrast, a high level of genetic stability and attenuation of classical cold-adapted attenuated viruses is controlled by a complex of 5-6 mutated internal genes. In other words, there is a multiple gene segment control of the attenuated phenotype of the influenza virus.

MOLECULAR MECHANISMS OF VIRULENCE

In this chapter, we discussed the role of individual mutant genes of cold-adapted MDVs in their attenuation. It is also of high importance to identify which genes determine virulence, pathogenicity, and the host range of mammals. The mechanisms responsible for a virus adaptation to a new host species or its enhanced virulence are not fully understood. There is a number of unsystematic studies in which practically every gene of the WT virus has found its place in the formation of virulence. The contribution of genes coding HA, PB2, NP, or NA proteins to enhanced pathogenicity and host adaptation was established in numerous studies [108 - 119].

An understanding of the determinants of pathogenicity and mechanisms of the airborne transmission pathway makes it possible to realistically assess the threat of zoonotic influenza viruses to human health. By now, knowledge about these numerous and interdependent determinants is at the stage of accumulating facts, and the molecular mechanisms of their functioning have yet to be established. The nature of determinants of host adaptation, specificity, pathogenicity, and transmissibility is polygenic, varies among influenza virus strains, depends on the host organism, and has different phenotypic manifestations [59, 120]. Host tropism and virulence are the results of molecular interactions between viral proteins and cellular factors. Influenza viruses use a variety of strategies to counter host antivirus programs. The manifestation of pathogenicity and transmissibility of viruses is a consequence of mutations and/or reassortment during adaptation to a new host species.

HA and NA as Determinants of Pathogenicity

The important contribution to the manifestation of tissue tropism, virulence, and interspecies transmission of influenza viruses belongs to the HA and the NA proteins functionally associated with it.

The pathogenicity of highly pathogenic avian influenza (HPAI) viruses of serotypes H5 and H7 is largely determined by the presence of a polybasic sequence in the cleavage site of the glycoprotein HA0 precursor, which makes HA susceptible to activation by ubiquitous cell proteases [121]. In low-virulent viruses, a single amino acid residue, arginine, is usually present at the cleavage site, which is well screened by carbohydrate chains and therefore less accessible to proteases. At the same time, the highly virulent 1918 Spanish flu virus lacks a polybasic HA cleavage site [122]. It has a different mechanism in which viral neuraminidase promotes more effective HA activation. NA of 1918 Spanish flu virus and some other early influenza viruses are able to bind host proteases with a fragment at the N-terminus of the structure. The enzyme bound to the virus, plasminogen, enters the endosome, where it is converted into plasmin, and already in the process of penetration into the cell cleaves HA [123].

It was also noted that the pathogenicity of avian influenza viruses for mammals could be realized through the shortening of the neuraminidase stalk [124].

The balance of HA and NA also plays an important role in efficient and sustainable transmissibility between humans; this was demonstrated by the example of the H1N1pdm09 virus, whose neuraminidase is characterized by a very high enzymatic activity [125].

Polymerase Complex as a Determinant of Pathogenicity and Transmissibility

The adaptation of the avian influenza virus to mammals leads to an increase in the activity of viral polymerase due to the interaction of polymerase with cellular nuclear import mechanisms. The RNA-dependent RNA polymerase of avian influenza viruses, containing the E627 in PB2 protein, does not work well in mammalian cells, since there are powerful factors that selectively limit the functions of foreign polymerases [126]. The replacement of E627L, as a determinant of the host range for mammals, has been described *in vitro* [127] and *in vivo* [128] in mice and guinea pigs (Table **4**). In guinea pigs, an increase in avian virus replication was noted at a temperature of 33 °C, which is probably due to the stable transmission to mammals [129].

Table 4. Role of Individual Genes and Mutations in Pathogenicity of Viruses.

Gene	Role of Individual Genes and Mutations in Pathogenicity of Viruses		
	H1N1 1918 Viruses	**HPAI H5 Viruses**	**LPAI/HPAI H7 Viruses**
HA	[123, 138]	[57, 140]	[141]
NA	[123]	No data	No data
PB2	E627K [59]	E627K [5]; D701N [142]	E627K [57, 141]; D701N, Q591K, M535L [131]
NS1	[135]	NS1 [143]	No data

This finding was supported by studies in ferrets, demonstrated that the replacement of L627E prevented the aerosol transmissibility of the 1918 Spanish flu virus [130]. Reconstruction of functional ribonucleoprotein *in vivo* showed that the polymerase complex of avian influenza A viruses is sensitive to the temperature of 33-37 °C. A decrease in the ability of the polymerase complex of avian viruses to replicate at 33 °C may be an obstacle to efficient growth in humans [13].

The PB2 E627L mutation has not always been required for human infection with highly pathogenic H5 and H7 avian influenza viruses. In such a situation, all HPAI H5 and H7 viruses had other substitutions in polymerase complex proteins involved in adaptation to replication in humans [131, 132].

The genomes of avian and human influenza A viruses differ in 52 positions associated with the host species, of which 35 differences are located in genes encoding polymerase complex [133]. The inclusion of a single PB2 gene of the HPAI A/Hong Kong/483/97 (H5N1) virus into the genome of LPAI A/Hong Kong/486/97 (H5N1) virus resulted in a lethal phenotype [5].

The PB1-F2 protein destabilizes mitochondria due to the creating pores in lipid membranes, which leads to leakage of macromolecules and ends with apoptosis [21, 22, 47, 48]. PB1-F2 also affects the activity of the polymerase complex by interacting with the PB1 protein [49]. All the activity of the PB1-F2 protein is aimed at an increase in the pathogenicity of the virus.

NS1 Protein as a Determinant of Pathogenicity

Escaping the host immune response is an undeniable determinant of pathogenicity. The innate immune system is the body's first line of defense. In this system, type I interferon plays a key role. The NS1 virus protein acts as an interferon antagonist and plays a multifunctional role in preventing the activation of interferon transcription factors [134]. The role of the NS1 protein as an interferon antagonist has been shown, in particular, in inbred mice lacking an Mx1 gene. This gene is an important factor in bringing cells into antiviral activity; the absence of the gene makes it impossible for the protective factors of cells to resist the virus [135]. The determinant of pathogenicity of the 1918 Spanish flu virus and H5N1 HPAI viruses probably is the insertion of four amino acids into its C-terminal sequence of the NS1 gene [46, 134, 135].

The 1918 H1N1 and H5N1 highly pathogenic viruses share the ability to induce abnormal innate immune responses in mice and non-human primates [136, 137]. This property is likely to be the cause of the pathogenicity of the Spanish flu virus and the high mortality rate in humans [132]. Studies of the Spanish flu virus recreated by reverse genetics methods found that the HA protein [138], the replicative complex [139], the NS1 [135] and PB1-F2 [23] contribute to its virulence, while the HA and PB2 proteins determine transmissibility [130].

Thus, all these data indicate that the determinants of pathogenicity of influenza viruses for mammals and humans were found in the number of proteins and amino acid positions of highly pathogenic zoonotic and pandemic influenza viruses (Table **4**).

ARE ATTENUATION AND PATHOGENICITY OF THE INFLUENZA VIRUS TWO SIDES OF THE SAME COIN?

Despite the abundance of publications, it is difficult and most likely impossible to systematize the information. Practically for each studied virus, its determinants of pathogenicity were found. According to various authors, almost every gene is involved in the formation of pathogenicity.

In the first part of this chapter, when we discussed the role of individual mutant genes of cold-adapted MDVs in their attenuation, we analyzed what will happen if replace individual WT genes of the WT virus with mutant genes of MDV. Therefore, trying to understand whether there is some universal mechanism for the formation of pathogenicity, we came the same way and systematized the knowledge of what will happen if a single "cold" gene of attenuated cold-adapted MDV is replaced with the gene of a WT virus, to estimate when the reversion to the WT phenotype will occur. We did this on the example of L17 MDV (Fig. **3**, right panel).

It seems to us that it would be logical to assume that if the inclusion of a certain "cold" gene in the genome of a WT virus leads to its attenuation, and the inclusion of the same WT gene in the genome of a "cold" virus leads to the reversion of its phenotype to the WT, then this gene plays a unique leading role in attenuation/virulence. In light of this, everything falls into place. This approach does not limit the scope of research to the results of studying a particular virus and may, we believe, lead us to an understanding of the universal mechanisms of attenuation/virulence that certainly exist. Thus, it is probably more correct to say that both attenuation and pathogenicity of the influenza virus are two sides of the same coin (Fig. **4**), or in other words, attenuation is the reverse side of pathogenicity. In this case, numerous disparate data are ordered and fall into their strictly defined place if they do not separate attenuation and pathogenicity.

Fig. (4). Attenuation and pathogenicity of the influenza virus could be considered as two sides of the same coin.

Following this point of view, we may conclude that polymerase genes play a unique role in the manifestation of certain biological properties of the influenza virus, in particular, such important attributes as its pathogenicity or attenuation.

CONCLUDING REMARKS

Advances of molecular genetics in the study of the influenza virus genome structure made it possible to decipher the main stages of replication and determine the mechanisms of the influenza virus attenuation. The genes that are responsible for attenuation were identified. The polymerase gene segments independently controlled the manifestation of the *ts*/att phenotype of both A and B type influenza viruses. The other genes coding for internal proteins played no crucial role in this respect.

REFERENCES

[1] Shi Y, Wu Y, Zhang W, Qi J, Gao GF. Enabling the 'host jump': structural determinants of receptor-binding specificity in influenza A viruses. Nat Rev Microbiol 2014; 12(12): 822-31. [http://dx.doi.org/10.1038/nrmicro3362] [PMID: 25383601]

[2] Wright PF, Neumann G, Kawaoka Y. Orthomyxoviruses. In: Knipe DM, Howley PM, Eds. Fields Virology. 5th ed. Philadelphia: Wolters Kluwer Health/Lippincott, Williams and Wilkins 2007; pp. 1691-740.

[3] Norkin LC. Virology: Molecular biology and pathogenesis. 1st ed., Washington, DC: ASM Press 2010. [http://dx.doi.org/10.1128/9781555814533]

[4] Imai M, Kawasaki K, Odagiri T. Cytoplasmic domain of influenza B virus BM2 protein plays critical roles in production of infectious virus. J Virol 2008; 82(2): 728-39. [http://dx.doi.org/10.1128/JVI.01752-07] [PMID: 17989175]

[5] Hatta M, Gao P, Halfmann P, Kawaoka Y. Molecular basis for high virulence of Hong Kong H5N1 influenza A viruses. Science 2001; 293(5536): 1840-2. [http://dx.doi.org/10.1126/science.1062882] [PMID: 11546875]

[6] Yasuda J, Nakada S, Kato A, Toyoda T, Ishihama A. Molecular assembly of influenza virus: association of the NS2 protein with virion matrix. Virology 1993; 196(1): 249-55. [http://dx.doi.org/10.1006/viro.1993.1473] [PMID: 8356796]

[7] O'Neill RE, Talon J, Palese P. The influenza virus NEP (NS2 protein) mediates the nuclear export of viral ribonucleoproteins. EMBO J 1998; 17(1): 288-96. [http://dx.doi.org/10.1093/emboj/17.1.288] [PMID: 9427762]

[8] Paragas J, Talon J, O'Neill RE, Anderson DK, García-Sastre A, Palese P. Influenza B and C virus NEP (NS2) proteins possess nuclear export activities. J Virol 2001; 75(16): 7375-83. [http://dx.doi.org/10.1128/JVI.75.16.7375-7383.2001] [PMID: 11462009]

[9] Paterson D, Fodor E. Emerging roles for the influenza A virus nuclear export protein (NEP). PLoS Pathog 2012; 8(12): e1003019. [http://dx.doi.org/10.1371/journal.ppat.1003019] [PMID: 23236273]

[10] Braam J, Ulmanen I, Krug RM. Molecular model of a eucaryotic transcription complex: functions and movements of influenza P proteins during capped RNA-primed transcription. Cell 1983; 34(2): 609-18. [http://dx.doi.org/10.1016/0092-8674(83)90393-8] [PMID: 6616622]

[11] Dalton RM, Mullin AE, Amorim MJ, Medcalf E, Tiley LS, Digard P. Temperature sensitive influenza A virus genome replication results from low thermal stability of polymerase-cRNA complexes. Virol J 2006; 3: 58. [http://dx.doi.org/10.1186/1743-422X-3-58] [PMID: 16934156]

[12] Kashiwagi T, Hara K, Nakazono Y, Hamada N, Watanabe H. Artificial hybrids of influenza A virus

RNA polymerase reveal PA subunit modulates its thermal sensitivity. PLoS One 2010; 5(12): e15140. [http://dx.doi.org/10.1371/journal.pone.0015140] [PMID: 21151876]

[13] Massin P, van der Werf S, Naffakh N. Residue 627 of PB2 is a determinant of cold sensitivity in RNA replication of avian influenza viruses. J Virol 2001; 75(11): 5398-404. [http://dx.doi.org/10.1128/JVI.75.11.5398-5404.2001] [PMID: 11333924]

[14] Nakatsu S, Murakami S, Shindo K, *et al.* Influenza C and D viruses package eight organized ribonucleoprotein complexes. J Virol 2018; 92(6): e02084-17. [http://dx.doi.org/10.1128/JVI.02084-17] [PMID: 29321324]

[15] Rosenthal PB, Zhang X, Formanowski F, *et al.* Structure of the haemagglutinin-esterase-fusion glycoprotein of influenza C virus. Nature 1998; 396(6706): 92-6. [http://dx.doi.org/10.1038/23974] [PMID: 9817207]

[16] Sugita Y, Sagara H, Noda T, Kawaoka Y. Configuration of viral ribonucleoprotein complexes within the influenza A virion. J Virol 2013; 87(23): 12879-84. [http://dx.doi.org/10.1128/JVI.02096-13] [PMID: 24067952]

[17] Baigent SJ, McCauley JW. Influenza type A in humans, mammals and birds: determinants of virus virulence, host-range and interspecies transmission. BioEssays 2003; 25(7): 657-71. [http://dx.doi.org/10.1002/bies.10303] [PMID: 12815721]

[18] Yewdell J, García-Sastre A. Influenza virus still surprises. Curr Opin Microbiol 2002; 5(4): 414-8. [http://dx.doi.org/10.1016/S1369-5274(02)00346-6] [PMID: 12160862]

[19] Krug RM, Fodor E. The virus genome and its replication.Textbook of influenza. 2nd ed. UK: Wiley Blackwell 2013; pp. 57-66. [http://dx.doi.org/10.1002/9781118636817.ch4]

[20] Bouvier NM, Palese P. The biology of influenza viruses. Vaccine 2008; 26(Suppl 4): D49-53. [http://dx.doi.org/10.1016/j.vaccine.2008.07.039] [PMID: 19230160]

[21] Chen W, Calvo PA, Malide D, *et al.* A novel influenza A virus mitochondrial protein that induces cell death. Nat Med 2001; 7(12): 1306-12. [http://dx.doi.org/10.1038/nm1201-1306] [PMID: 11726970]

[22] Zamarin D, García-Sastre A, Xiao X, Wang R, Palese P. Influenza virus PB1-F2 protein induces cell death through mitochondrial ANT3 and VDAC1. PLoS Pathog 2005; 1(1): e4. [http://dx.doi.org/10.1371/journal.ppat.0010004] [PMID: 16201016]

[23] McAuley JL, Hornung F, Boyd KL, *et al.* Expression of the 1918 influenza A virus PB1-F2 enhances the pathogenesis of viral and secondary bacterial pneumonia. Cell Host Microbe 2007; 2(4): 240-9. [http://dx.doi.org/10.1016/j.chom.2007.09.001] [PMID: 18005742]

[24] Tauber S, Ligertwood Y, Quigg-Nicol M, Dutia BM, Elliott RM. Behaviour of influenza A viruses differentially expressing segment 2 gene products *in vitro* and *in vivo*. J Gen Virol 2012; 93(Pt 4): 840-9. [http://dx.doi.org/10.1099/vir.0.039966-0] [PMID: 22190016]

[25] Wise HM, Hutchinson EC, Jagger BW, *et al.* Identification of a novel splice variant form of the influenza A virus M2 ion channel with an antigenically distinct ectodomain. PLoS Pathog 2012; 8(11): e1002998. [http://dx.doi.org/10.1371/journal.ppat.1002998] [PMID: 23133386]

[26] Jagger BW, Wise HM, Kash JC, *et al.* An overlapping protein-coding region in influenza A virus segment 3 modulates the host response. Science 2012; 337(6091): 199-204. [http://dx.doi.org/10.1126/science.1222213] [PMID: 22745253]

[27] Wagner R, Matrosovich M, Klenk HD. Functional balance between haemagglutinin and neuraminidase in influenza virus infections. Rev Med Virol 2002; 12(3): 159-66. [http://dx.doi.org/10.1002/rmv.352] [PMID: 11987141]

[28] Matrosovich MN, Matrosovich TY, Gray T, Roberts NA, Klenk HD. Neuraminidase is important for the initiation of influenza virus infection in human airway epithelium. J Virol 2004; 78(22): 12665-7. [http://dx.doi.org/10.1128/JVI.78.22.12665-12667.2004] [PMID: 15507653]

[29] Nayak DP, Hui EK, Barman S. Assembly and budding of influenza virus. Virus Res 2004; 106(2): 147-65. [http://dx.doi.org/10.1016/j.virusres.2004.08.012] [PMID: 15567494]

[30] Hatta M, Kawaoka Y. The NB protein of influenza B virus is not necessary for virus replication *in vitro*. J Virol 2003; 77(10): 6050-4. [http://dx.doi.org/10.1128/JVI.77.10.6050-6054.2003] [PMID: 12719596]

[31] Liu M, Lam MK, Zhang Q, Elderfield R, Barclay WS, Shaw PC. The functional study of the N-terminal region of influenza B virus nucleoprotein. PLoS One 2015; 10(9): e0137802. [http://dx.doi.org/10.1371/journal.pone.0137802] [PMID: 26368391]

[32] Ng AK, Lam MK, Zhang H, *et al.* Structural basis for RNA binding and homo-oligomer formation by influenza B virus nucleoprotein. J Virol 2012; 86(12): 6758-67. [http://dx.doi.org/10.1128/JVI.00073-12] [PMID: 22496219]

[33] Sherry L, Smith M, Davidson S, Jackson D. The N terminus of the influenza B virus nucleoprotein is essential for virus viability, nuclear localization, and optimal transcription and replication of the viral genome. J Virol 2014; 88(21): 12326-38. [http://dx.doi.org/10.1128/JVI.01542-14] [PMID: 25122787]

[34] Benton DJ, Gamblin SJ, Rosenthal PB, Skehel JJ. Structural transitions in influenza haemagglutinin at membrane fusion pH. Nature 2020; 583(7814): 150-3. [http://dx.doi.org/10.1038/s41586-020-2333-6] [PMID: 32461688]

[35] Steinhauer DA. Role of hemagglutinin cleavage for the pathogenicity of influenza virus. Virology 1999; 258(1): 1-20. [http://dx.doi.org/10.1006/viro.1999.9716] [PMID: 10329563]

[36] Tate MD, Job ER, Deng YM, Gunalan V, Maurer-Stroh S, Reading PC. Playing hide and seek: how glycosylation of the influenza virus hemagglutinin can modulate the immune response to infection. Viruses 2014; 6(3): 1294-316. [http://dx.doi.org/10.3390/v6031294] [PMID: 24638204]

[37] Gallagher PJ, Henneberry JM, Sambrook JF, Gething MJ. Glycosylation requirements for intracellular transport and function of the hemagglutinin of influenza virus. J Virol 1992; 66(12): 7136-45. [http://dx.doi.org/10.1128/JVI.66.12.7136-7145.1992] [PMID: 1331514]

[38] Kim JI, Park MS. N-linked glycosylation in the hemagglutinin of influenza A viruses. Yonsei Med J 2012; 53(5): 886-93. [http://dx.doi.org/10.3349/ymj.2012.53.5.886] [PMID: 22869469]

[39] Chen BJ, Takeda M, Lamb RA. Influenza virus hemagglutinin (H3 subtype) requires palmitoylation of its cytoplasmic tail for assembly: M1 proteins of two subtypes differ in their ability to support assembly. J Virol 2005; 79(21): 13673-84. [http://dx.doi.org/10.1128/JVI.79.21.13673-13684.2005] [PMID: 16227287]

[40] Skehel JJ, Wiley DC. Receptor binding and membrane fusion in virus entry: the influenza hemagglutinin. Annu Rev Biochem 2000; 69: 531-69. [http://dx.doi.org/10.1146/annurev.biochem.69.1.531] [PMID: 10966468]

[41] Hampson AW, Mackenzie JS. The influenza viruses. Med J Aust 2006; 185(S10): S39-43. [http://dx.doi.org/10.5694/j.1326-5377.2006.tb00705.x] [PMID: 17115950]

[42] Wang Q. Influenza type B virus haemagglutinin: antigenicity, receptor binding and membrane fusion. In: Wang Q, Tao YJ, Eds. Influenza: Molecular Virology. UK: Caister Academic Press 2010; p. 197.

[43] Gamblin SJ, Skehel JJ. Influenza hemagglutinin and neuraminidase membrane glycoproteins. J Biol

Chem 2010; 285(37): 28403-9.
[http://dx.doi.org/10.1074/jbc.R110.129809] [PMID: 20538598]

[44] Shtyrya YA, Mochalova LV, Bovin NV. Influenza virus neuraminidase: structure and function. Acta Naturae 2009; 1(2): 26-32.
[http://dx.doi.org/10.32607/20758251-2009-1-2-26-32] [PMID: 22649600]

[45] Wise HM, Barbezange C, Jagger BW, *et al.* Overlapping signals for translational regulation and packaging of influenza A virus segment 2. Nucleic Acids Res 2011; 39(17): 7775-90.
[http://dx.doi.org/10.1093/nar/gkr487] [PMID: 21693560]

[46] Jackson D, Elderfield RA, Barclay WS. Molecular studies of influenza B virus in the reverse genetics era. J Gen Virol 2011; 92(Pt 1): 1-17.
[http://dx.doi.org/10.1099/vir.0.026187-0] [PMID: 20926635]

[47] Chanturiya AN, Basañez G, Schubert U, Henklein P, Yewdell JW, Zimmerberg J. PB1-F2, an influenza A virus-encoded proapoptotic mitochondrial protein, creates variably sized pores in planar lipid membranes. J Virol 2004; 78(12): 6304-12.
[http://dx.doi.org/10.1128/JVI.78.12.6304-6312.2004] [PMID: 15163724]

[48] Henkel M, Mitzner D, Henklein P, *et al.* The proapoptotic influenza A virus protein PB1-F2 forms a nonselective ion channel. PLoS One 2010; 5(6): e11112.
[http://dx.doi.org/10.1371/journal.pone.0011112] [PMID: 20559552]

[49] Conenello GM, Tisoncik JR, Rosenzweig E, Varga ZT, Palese P, Katze MG. A single N66S mutation in the PB1-F2 protein of influenza A virus increases virulence by inhibiting the early interferon response *in vivo*. J Virol 2011; 85(2): 652-62.
[http://dx.doi.org/10.1128/JVI.01987-10] [PMID: 21084483]

[50] Nayak D, Shivakoti S, Balogun RA, Lee G, Zhou ZH. Structure, disassembly, assembly, and budding of influenza viruses.Textbook of influenza. 2nd ed. UK: Wiley Blackwell 2013; pp. 37-56.
[http://dx.doi.org/10.1002/9781118636817.ch3]

[51] Fodor E, Te Velthuis AJW. Structure and function of the influenza virus transcription and replication machinery. Cold Spring Harb Perspect Med 2020; 10(9): a038398.
[http://dx.doi.org/10.1101/cshperspect.a038398] [PMID: 31871230]

[52] Connor RJ, Kawaoka Y, Webster RG, Paulson JC. Receptor specificity in human, avian, and equine H2 and H3 influenza virus isolates. Virology 1994; 205(1): 17-23.
[http://dx.doi.org/10.1006/viro.1994.1615] [PMID: 7975212]

[53] Gambaryan AS, Tuzikov AB, Piskarev VE, *et al.* Specification of receptor-binding phenotypes of influenza virus isolates from different hosts using synthetic sialylglycopolymers: non-egg-adapted human H1 and H3 influenza A and influenza B viruses share a common high binding affinity for 6′-sialyl(N-acetyllactosamine). Virology 1997; 232(2): 345-50.
[http://dx.doi.org/10.1006/viro.1997.8572] [PMID: 9191848]

[54] Matrosovich M, Tuzikov A, Bovin N, *et al.* Early alterations of the receptor-binding properties of H1, H2, and H3 avian influenza virus hemagglutinins after their introduction into mammals. J Virol 2000; 74(18): 8502-12.
[http://dx.doi.org/10.1128/JVI.74.18.8502-8512.2000] [PMID: 10954551]

[55] Rogers GN, Pritchett TJ, Lane JL, Paulson JC. Differential sensitivity of human, avian, and equine influenza A viruses to a glycoprotein inhibitor of infection: selection of receptor specific variants. Virology 1983; 131(2): 394-408.
[http://dx.doi.org/10.1016/0042-6822(83)90507-X] [PMID: 6197808]

[56] Couceiro JN, Paulson JC, Baum LG. Influenza virus strains selectively recognize sialyloligosaccharides on human respiratory epithelium; the role of the host cell in selection of hemagglutinin receptor specificity. Virus Res 1993; 29(2): 155-65.
[http://dx.doi.org/10.1016/0168-1702(93)90056-S] [PMID: 8212857]

[57] Shinya K, Ebina M, Yamada S, Ono M, Kasai N, Kawaoka Y. Avian flu: influenza virus receptors in the human airway. Nature 2006; 440(7083): 435-6. [http://dx.doi.org/10.1038/440435a] [PMID: 16554799]

[58] Matrosovich MN, Matrosovich TY, Gray T, Roberts NA, Klenk HD. Human and avian influenza viruses target different cell types in cultures of human airway epithelium. Proc Natl Acad Sci USA 2004; 101(13): 4620-4. [http://dx.doi.org/10.1073/pnas.0308001101] [PMID: 15070767]

[59] Taubenberger JK, Kash JC. Influenza virus evolution, host adaptation, and pandemic formation. Cell Host Microbe 2010; 7(6): 440-51. [http://dx.doi.org/10.1016/j.chom.2010.05.009] [PMID: 20542248]

[60] Webby R, Richt J. Influenza in swine.Textbook of influenza. 2nd ed. UK: Wiley Blackwell 2013; pp. 190-202. [http://dx.doi.org/10.1002/9781118636817.ch12]

[61] Wilks S, de Graaf M, Smith DJ, Burke DF. A review of influenza haemagglutinin receptor binding as it relates to pandemic properties. Vaccine 2012; 30(29): 4369-76. [http://dx.doi.org/10.1016/j.vaccine.2012.02.076] [PMID: 22682293]

[62] Wang YF, Chang CF, Chi CY, Wang HC, Wang JR, Su IJ. Characterization of glycan binding specificities of influenza B viruses with correlation with hemagglutinin genotypes and clinical features. J Med Virol 2012; 84(4): 679-85. [http://dx.doi.org/10.1002/jmv.23219] [PMID: 22337309]

[63] Lakadamyali M, Rust MJ, Babcock HP, Zhuang X. Visualizing infection of individual influenza viruses. Proc Natl Acad Sci USA 2003; 100(16): 9280-5. [http://dx.doi.org/10.1073/pnas.0832269100] [PMID: 12883000]

[64] Gao J, Gui M, Xiang Y. Structural intermediates in the low pH-induced transition of influenza hemagglutinin. PLoS Pathog 2020; 16(11): e1009062. [http://dx.doi.org/10.1371/journal.ppat.1009062] [PMID: 33253316]

[65] Boriskin YS, Leneva IA, Pécheur EI, Polyak SJ. Arbidol: a broad-spectrum antiviral compound that blocks viral fusion. Curr Med Chem 2008; 15(10): 997-1005. [http://dx.doi.org/10.2174/092986708784049658] [PMID: 18393857]

[66] Kadam RU, Wilson IA. Structural basis of influenza virus fusion inhibition by the antiviral drug Arbidol. Proc Natl Acad Sci USA 2017; 114(2): 206-14. [http://dx.doi.org/10.1073/pnas.1617020114] [PMID: 28003465]

[67] Daidoji T, Watanabe Y, Arai Y, Kajikawa J, Hirose R, Nakaya T. Unique infectious strategy of H5N1 avian influenza virus is governed by the acid-destabilized property of hemagglutinin. Viral Immunol 2017; 30(6): 398-407. [http://dx.doi.org/10.1089/vim.2017.0020] [PMID: 28654310]

[68] Martin K, Helenius A. Transport of incoming influenza virus nucleocapsids into the nucleus. J Virol 1991; 65(1): 232-44. [http://dx.doi.org/10.1128/JVI.65.1.232-244.1991] [PMID: 1985199]

[69] Cros JF, Palese P. Trafficking of viral genomic RNA into and out of the nucleus: influenza, Thogoto and Borna disease viruses. Virus Res 2003; 95(1-2): 3-12. [http://dx.doi.org/10.1016/S0168-1702(03)00159-X] [PMID: 12921991]

[70] Matsuoka Y, Matsumae H, Katoh M, *et al.* A comprehensive map of the influenza A virus replication cycle. BMC Syst Biol 2013; 7: 97. [http://dx.doi.org/10.1186/1752-0509-7-97] [PMID: 24088197]

[71] Neumann G, Brownlee GG, Fodor E, Kawaoka Y. Orthomyxovirus replication, transcription, and polyadenylation. Curr Top Microbiol Immunol 2004; 283: 121-43. [http://dx.doi.org/10.1007/978-3-662-06099-5_4] [PMID: 15298169]

[72] Li X, Palese P. Characterization of the polyadenylation signal of influenza virus RNA. J Virol 1994; 68(2): 1245-9.
[http://dx.doi.org/10.1128/jvi.68.2.1245-1249.1994] [PMID: 7507179]

[73] Bier K, York A, Fodor E. Cellular cap-binding proteins associate with influenza virus mRNAs. J Gen Virol 2011; 92(Pt 7): 1627-34.
[http://dx.doi.org/10.1099/vir.0.029231-0] [PMID: 21402597]

[74] Yángüez E, Rodriguez P, Goodfellow I, Nieto A. Influenza virus polymerase confers independence of the cellular cap-binding factor eIF4E for viral mRNA translation. Virology 2012; 422(2): 297-307.
[http://dx.doi.org/10.1016/j.virol.2011.10.028] [PMID: 22112850]

[75] Schneider J, Wolff T. Nuclear functions of the influenza A and B viruses NS1 proteins: do they play a role in viral mRNA export? Vaccine 2009; 27(45): 6312-6.
[http://dx.doi.org/10.1016/j.vaccine.2009.01.015] [PMID: 19840666]

[76] Webster RG, Rott R. Influenza virus A pathogenicity: the pivotal role of hemagglutinin. Cell 1987; 50(5): 665-6.
[http://dx.doi.org/10.1016/0092-8674(87)90321-7] [PMID: 3304656]

[77] Horimoto T, Kawaoka Y. Reverse genetics provides direct evidence for a correlation of hemagglutinin cleavability and virulence of an avian influenza A virus. J Virol 1994; 68(5): 3120-8.
[http://dx.doi.org/10.1128/jvi.68.5.3120-3128.1994] [PMID: 8151777]

[78] Scheiffele P, Roth MG, Simons K. Interaction of influenza virus haemagglutinin with sphingolipid-cholesterol membrane domains *via* its transmembrane domain. EMBO J 1997; 16(18): 5501-8.
[http://dx.doi.org/10.1093/emboj/16.18.5501] [PMID: 9312009]

[79] Ruigrok RW, Barge A, Durrer P, Brunner J, Ma K, Whittaker GR. Membrane interaction of influenza virus M1 protein. Virology 2000; 267(2): 289-98.
[http://dx.doi.org/10.1006/viro.1999.0134] [PMID: 10662624]

[80] McCown MF, Pekosz A. The influenza A virus M2 cytoplasmic tail is required for infectious virus production and efficient genome packaging. J Virol 2005; 79(6): 3595-605.
[http://dx.doi.org/10.1128/JVI.79.6.3595-3605.2005] [PMID: 15731254]

[81] Fujii Y, Goto H, Watanabe T, Yoshida T, Kawaoka Y. Selective incorporation of influenza virus RNA segments into virions. Proc Natl Acad Sci USA 2003; 100(4): 2002-7.
[http://dx.doi.org/10.1073/pnas.0437772100] [PMID: 12574509]

[82] Hutchinson EC, von Kirchbach JC, Gog JR, Digard P. Genome packaging in influenza A virus. J Gen Virol 2010; 91(Pt 2): 313-28.
[http://dx.doi.org/10.1099/vir.0.017608-0] [PMID: 19955561]

[83] Noda T, Kawaoka Y. Packaging of influenza virus genome: robustness of selection. Proc Natl Acad Sci USA 2012; 109(23): 8797-8.
[http://dx.doi.org/10.1073/pnas.1206736109] [PMID: 22615368]

[84] Haralampiev I, Prisner S, Nitzan M, *et al.* Selective flexible packaging pathways of the segmented genome of influenza A virus. Nat Commun 2020; 11(1): 4355.
[http://dx.doi.org/10.1038/s41467-020-18108-1] [PMID: 32859915]

[85] WHO. Global action plan to increase vaccine supply for influenza vaccines. 2006, Geneva, Belgium. [Available from: http://whqlibdoc.who.int/hq/2006/WHO_IVB_06.13_eng.pdf]

[86] Rudenko LG, Slepushkin AN, Monto AS, *et al.* Efficacy of live attenuated and inactivated influenza vaccines in schoolchildren and their unvaccinated contacts in Novgorod, Russia. J Infect Dis 1993; 168(4): 881-7.
[http://dx.doi.org/10.1093/infdis/168.4.881] [PMID: 8376833]

[87] Alexandrova GI, Smorodintsev AA. Obtaining of an additionally attenuated vaccinating cryophilic influenza strain. Rev Roum Inframicrobiol 1965; 2: 179-89.

[88] Maassab HF. Adaptation and growth characteristics of influenza virus at 25 degrees c. Nature 1967; 213(5076): 612-4.
[http://dx.doi.org/10.1038/213612a0] [PMID: 6040602]

[89] Donabedian AM, DeBorde DC, Cook S, Smitka CW, Maassab HF. A mutation in the PA protein gene of cold-adapted B/Ann Arbor/1/66 influenza virus associated with reversion of temperature sensitivity and attenuated virulence. Virology 1988; 163(2): 444-51.
[http://dx.doi.org/10.1016/0042-6822(88)90285-1] [PMID: 3354203]

[90] Odagiri T, Tosaka A, Ishida N, Maassab HF. Biological characteristics of a cold-adapted influenza A virus mutation residing on a polymerase gene. Arch Virol 1986; 88(1-2): 91-104.
[http://dx.doi.org/10.1007/BF01310893] [PMID: 2420313]

[91] Snyder MH, Betts RF, DeBorde D, *et al.* Four viral genes independently contribute to attenuation of live influenza A/Ann Arbor/6/60 (H2N2) cold-adapted reassortant virus vaccines. J Virol 1988; 62(2): 488-95.
[http://dx.doi.org/10.1128/JVI.62.2.488-495.1988] [PMID: 3336068]

[92] Klimov AI, Kiseleva IV, Alexandrova GI, Cox NJ. Genes coding for polymerase proteins are essential for attenuation of the cold–adapted A/Leningrad/134/17/57 (H2N2) influenza virus. In: proceedings of Options for the Control of Influenza IV; ADME Osterhaus, N Cox, AW Hampson, Eds. 2001; pp. 955-9. Crete, Greece, 23-28 September 2000: Elsevier.

[93] Kiseleva IV, Voeten JT, Teley LC, *et al.* PB2 and PA genes control the expression of the temperature-sensitive phenotype of cold-adapted B/USSR/60/69 influenza master donor virus. J Gen Virol 2010; 91(Pt 4): 931-7.
[http://dx.doi.org/10.1099/vir.0.017996-0] [PMID: 20007357]

[94] Kiseleva I, Klimov A, Su Q, Szymkowiak C, Toner TJ, Kwan W-S, Eds. Role of individual genes of the A/Leningrad/134/17/57 (H2N2) cold–adapted donor strain in manifestation of the temperature sensitive phenotype of reassortant influenza A viruses. In: Proceedings of Options for the Control of Influenza V, Y Kawaoka, ed. 2003; pp. 547-50. Okinawa, Japan, 6–9 October 2003: Elsevier.

[95] Jin H, Lu B, Zhou H, *et al.* Multiple amino acid residues confer temperature sensitivity to human influenza virus vaccine strains (FluMist) derived from cold-adapted A/Ann Arbor/6/60. Virology 2003; 306(1): 18-24.
[http://dx.doi.org/10.1016/S0042-6822(02)00035-1] [PMID: 12620793]

[96] Hoffmann E, Mahmood K, Chen Z, *et al.* Multiple gene segments control the temperature sensitivity and attenuation phenotypes of ca B/Ann Arbor/1/66. J Virol 2005; 79(17): 11014-21.
[http://dx.doi.org/10.1128/JVI.79.17.11014-11021.2005] [PMID: 16103152]

[97] Isakova-Sivak I, Chen LM, Matsuoka Y, *et al.* Genetic bases of the temperature-sensitive phenotype of a master donor virus used in live attenuated influenza vaccines: A/Leningrad/134/17/57 (H2N2). Virology 2011; 412(2): 297-305.
[http://dx.doi.org/10.1016/j.virol.2011.01.004] [PMID: 21315402]

[98] Klimov AI, Cox NJ, Yotov WV, Rocha E, Alexandrova GI, Kendal AP. Sequence changes in the live attenuated, cold-adapted variants of influenza A/Leningrad/134/57 (H2N2) virus. Virology 1992; 186(2): 795-7.
[http://dx.doi.org/10.1016/0042-6822(92)90050-Y] [PMID: 1733114]

[99] Jin H, Zhou H, Lu B, Kemble G. Imparting temperature sensitivity and attenuation in ferrets to A/Puerto Rico/8/34 influenza virus by transferring the genetic signature for temperature sensitivity from cold-adapted A/Ann Arbor/6/60. J Virol 2004; 78(2): 995-8.
[http://dx.doi.org/10.1128/JVI.78.2.995-998.2004] [PMID: 14694130]

[100] Odagiri T, Tanaka T, Tobita K. Temperature-sensitive defect of influenza A/Ann Arbor/6/60 cold-adapted variant leads to a blockage of matrix polypeptide incorporation into the plasma membrane of the infected cells. Virus Res 1987; 7(3): 203-18.
[http://dx.doi.org/10.1016/0168-1702(87)90028-1] [PMID: 3604455]

[101] McCauley JW, Penn CR. The critical cut-off temperature of avian influenza viruses. Virus Res 1990; 17(3): 191-8.
[http://dx.doi.org/10.1016/0168-1702(90)90064-I] [PMID: 2077784]

[102] Kiseleva IV, Larionova NV, Voeten JT, *et al.* Leading role of genes coding polymerase complex in attenuation of domestic donor viruses for A and B live influenza vaccine. Zh Mikrobiol Epidemiol Immunobiol 2010; (6): 41-7.
[PMID: 21381376]

[103] He W, Wang W, Han H, Wang L, Zhang G, Gao B. Molecular basis of live-attenuated influenza virus. PLoS One 2013; 8(3): e60413.
[http://dx.doi.org/10.1371/journal.pone.0060413] [PMID: 23555969]

[104] Romanova J, Katinger D, Ferko B, *et al.* Live cold-adapted influenza A vaccine produced in Vero cell line. Virus Res 2004; 103(1-2): 187-93.
[http://dx.doi.org/10.1016/j.virusres.2004.02.032] [PMID: 15163508]

[105] Richt JA, García-Sastre A. Attenuated influenza virus vaccines with modified NS1 proteins. Curr Top Microbiol Immunol 2009; 333: 177-95.
[http://dx.doi.org/10.1007/978-3-540-92165-3_9] [PMID: 19768406]

[106] Wressnigg N, Voss D, Wolff T, *et al.* Development of a live-attenuated influenza B DeltaNS1 intranasal vaccine candidate. Vaccine 2009; 27(21): 2851-7.
[http://dx.doi.org/10.1016/j.vaccine.2009.02.087] [PMID: 19366569]

[107] Wacheck V, Egorov A, Groiss F, *et al.* A novel type of influenza vaccine: safety and immunogenicity of replication-deficient influenza virus created by deletion of the interferon antagonist NS1. J Infect Dis 2010; 201(3): 354-62.
[http://dx.doi.org/10.1086/649428] [PMID: 20039806]

[108] Choi EJ, Lee YJ, Lee JM, *et al.* The effect of mutations derived from mouse-adapted H3N2 seasonal influenza A virus to pathogenicity and host adaptation. PLoS One 2020; 15(1): e0227516.
[http://dx.doi.org/10.1371/journal.pone.0227516] [PMID: 31917822]

[109] Gabriel G, Czudai-Matwich V, Klenk HD. Adaptive mutations in the H5N1 polymerase complex. Virus Res 2013; 178(1): 53-62.
[http://dx.doi.org/10.1016/j.virusres.2013.05.010] [PMID: 23732876]

[110] Gabriel G, Fodor E. Molecular determinants of pathogenicity in the polymerase complex. Curr Top Microbiol Immunol 2014; 385: 35-60.
[http://dx.doi.org/10.1007/82_2014_386] [PMID: 25033751]

[111] Hu M, Yuan S, Zhang K, *et al.* PB2 substitutions V598T/I increase the virulence of H7N9 influenza A virus in mammals. Virology 2017; 501: 92-101.
[http://dx.doi.org/10.1016/j.virol.2016.11.008] [PMID: 27889648]

[112] Mänz B, Schwemmle M, Brunotte L. Adaptation of avian influenza A virus polymerase in mammals to overcome the host species barrier. J Virol 2013; 87(13): 7200-9.
[http://dx.doi.org/10.1128/JVI.00980-13] [PMID: 23616660]

[113] Mok CK, Lee HH, Lestra M, *et al.* Amino acid substitutions in polymerase basic protein 2 gene contribute to the pathogenicity of the novel A/H7N9 influenza virus in mammalian hosts. J Virol 2014; 88(6): 3568-76.
[http://dx.doi.org/10.1128/JVI.02740-13] [PMID: 24403592]

[114] Ping J, Dankar SK, Forbes NE, *et al.* PB2 and hemagglutinin mutations are major determinants of host range and virulence in mouse-adapted influenza A virus. J Virol 2010; 84(20): 10606-18.
[http://dx.doi.org/10.1128/JVI.01187-10] [PMID: 20702632]

[115] Rodriguez-Frandsen A, Alfonso R, Nieto A. Influenza virus polymerase: Functions on host range, inhibition of cellular response to infection and pathogenicity. Virus Res 2015; 209: 23-38.
[http://dx.doi.org/10.1016/j.virusres.2015.03.017] [PMID: 25916175]

[116] Song H, Qi J, Khedri Z, *et al.* An open receptor-binding cavity of hemagglutinin-esterase-fusion glycoprotein from newly-identified influenza D virus: Basis for its broad cell tropism. PLoS Pathog 2016; 12(1): e1005411.
[http://dx.doi.org/10.1371/journal.ppat.1005411] [PMID: 26816272]

[117] Tan L, Su S, Smith DK, *et al.* A combination of HA and PA mutations enhances virulence in a mouse-adapted H6N6 influenza A virus. J Virol 2014; 88(24): 14116-25.
[http://dx.doi.org/10.1128/JVI.01736-14] [PMID: 25275121]

[118] Watanabe T, Tisoncik-Go J, Tchitchek N, *et al.* 1918 Influenza virus hemagglutinin (HA) and the viral RNA polymerase complex enhance viral pathogenicity, but only HA induces aberrant host responses in mice. J Virol 2013; 87(9): 5239-54.
[http://dx.doi.org/10.1128/JVI.02753-12] [PMID: 23449804]

[119] Yamayoshi S, Yamada S, Fukuyama S, *et al.* Virulence-affecting amino acid changes in the PA protein of H7N9 influenza A viruses. J Virol 2014; 88(6): 3127-34.
[http://dx.doi.org/10.1128/JVI.03155-13] [PMID: 24371069]

[120] Böttcher-Friebertshäuser E, Garten W, Matrosovich M, Klenk HD. The hemagglutinin: a determinant of pathogenicity. Curr Top Microbiol Immunol 2014; 385: 3-34.
[http://dx.doi.org/10.1007/82_2014_384] [PMID: 25031010]

[121] Zambon MC. The pathogenesis of influenza in humans. Rev Med Virol 2001; 11(4): 227-41.
[http://dx.doi.org/10.1002/rmv.319] [PMID: 11479929]

[122] Reid AH, Taubenberger JK, Fanning TG. Evidence of an absence: the genetic origins of the 1918 pandemic influenza virus. Nat Rev Microbiol 2004; 2(11): 909-14.
[http://dx.doi.org/10.1038/nrmicro1027] [PMID: 15494747]

[123] Tumpey TM, Basler CF, Aguilar PV, *et al.* Characterization of the reconstructed 1918 Spanish influenza pandemic virus. Science 2005; 310(5745): 77-80.
[http://dx.doi.org/10.1126/science.1119392] [PMID: 16210530]

[124] Matsuoka Y, Swayne DE, Thomas C, *et al.* Neuraminidase stalk length and additional glycosylation of the hemagglutinin influence the virulence of influenza H5N1 viruses for mice. J Virol 2009; 83(9): 4704-8.
[http://dx.doi.org/10.1128/JVI.01987-08] [PMID: 19225004]

[125] Yen HL, Liang CH, Wu CY, *et al.* Hemagglutinin-neuraminidase balance confers respiratory-droplet transmissibility of the pandemic H1N1 influenza virus in ferrets. Proc Natl Acad Sci USA 2011; 108(34): 14264-9.
[http://dx.doi.org/10.1073/pnas.1111000108] [PMID: 21825167]

[126] Mehle A. Unusual influenza A viruses in bats. Viruses 2014; 6(9): 3438-49.
[http://dx.doi.org/10.3390/v6093438] [PMID: 25256392]

[127] Subbarao EK, London W, Murphy BR. A single amino acid in the PB2 gene of influenza A virus is a determinant of host range. J Virol 1993; 67(4): 1761-4.
[http://dx.doi.org/10.1128/JVI.67.4.1761-1764.1993] [PMID: 8445709]

[128] Maines TR, Jayaraman A, Belser JA, *et al.* Transmission and pathogenesis of swine-origin 2009 A(H1N1) influenza viruses in ferrets and mice. Science 2009; 325(5939): 484-7.
[http://dx.doi.org/10.1126/science.1177238] [PMID: 19574347]

[129] Lowen AC, Bouvier NM, Steel J. Transmission in the guinea pig model. Curr Top Microbiol Immunol 2014; 385: 157-83.
[http://dx.doi.org/10.1007/82_2014_390] [PMID: 25001209]

[130] Van Hoeven N, Pappas C, Belser JA, *et al.* Human HA and polymerase subunit PB2 proteins confer transmission of an avian influenza virus through the air. Proc Natl Acad Sci USA 2009; 106(9): 3366-71.
[http://dx.doi.org/10.1073/pnas.0813172106] [PMID: 19211790]

[131] Chen GW, Kuo SM, Yang SL, *et al.* Genomic signatures for avian H7N9 viruses adapting to humans. PLoS One 2016; 11(2): e0148432.
[http://dx.doi.org/10.1371/journal.pone.0148432] [PMID: 26845764]

[132] de Jong MD, Simmons CP, Thanh TT, *et al.* Fatal outcome of human influenza A (H5N1) is associated with high viral load and hypercytokinemia. Nat Med 2006; 12(10): 1203-7.
[http://dx.doi.org/10.1038/nm1477] [PMID: 16964257]

[133] Chen GW, Chang SC, Mok CK, *et al.* Genomic signatures of human *versus* avian influenza A viruses. Emerg Infect Dis 2006; 12(9): 1353-60.
[http://dx.doi.org/10.3201/eid1209.060276] [PMID: 17073083]

[134] Hale BG, Randall RE, Ortín J, Jackson D. The multifunctional NS1 protein of influenza A viruses. J Gen Virol 2008; 89(Pt 10): 2359-76.
[http://dx.doi.org/10.1099/vir.0.2008/004606-0] [PMID: 18796704]

[135] Geiss GK, Salvatore M, Tumpey TM, *et al.* Cellular transcriptional profiling in influenza A virus-infected lung epithelial cells: the role of the nonstructural NS1 protein in the evasion of the host innate defense and its potential contribution to pandemic influenza. Proc Natl Acad Sci USA 2002; 99(16): 10736-41.
[http://dx.doi.org/10.1073/pnas.112338099] [PMID: 12149435]

[136] Kash JC, Tumpey TM, Proll SC, *et al.* Genomic analysis of increased host immune and cell death responses induced by 1918 influenza virus. Nature 2006; 443(7111): 578-81.
[http://dx.doi.org/10.1038/nature05181] [PMID: 17006449]

[137] Kobasa D, Jones SM, Shinya K, *et al.* Aberrant innate immune response in lethal infection of macaques with the 1918 influenza virus. Nature 2007; 445(7125): 319-23.
[http://dx.doi.org/10.1038/nature05495] [PMID: 17230189]

[138] Tumpey TM, García-Sastre A, Taubenberger JK, Palese P, Swayne DE, Basler CF. Pathogenicity and immunogenicity of influenza viruses with genes from the 1918 pandemic virus. Proc Natl Acad Sci USA 2004; 101(9): 3166-71.
[http://dx.doi.org/10.1073/pnas.0308391100] [PMID: 14963236]

[139] Watanabe T, Watanabe S, Shinya K, Kim JH, Hatta M, Kawaoka Y. Viral RNA polymerase complex promotes optimal growth of 1918 virus in the lower respiratory tract of ferrets. Proc Natl Acad Sci USA 2009; 106(2): 588-92.
[http://dx.doi.org/10.1073/pnas.0806959106] [PMID: 19114663]

[140] van Riel D, Munster VJ, de Wit E, *et al.* H5N1 virus attachment to lower respiratory tract. Science 2006; 312(5772): 399.
[http://dx.doi.org/10.1126/science.1125548] [PMID: 16556800]

[141] Munster VJ, de Wit E, van Riel D, *et al.* The molecular basis of the pathogenicity of the Dutch highly pathogenic human influenza A H7N7 viruses. J Infect Dis 2007; 196(2): 258-65.
[http://dx.doi.org/10.1086/518792] [PMID: 17570113]

[142] Li Z, Chen H, Jiao P, *et al.* Molecular basis of replication of duck H5N1 influenza viruses in a mammalian mouse model. J Virol 2005; 79(18): 12058-64.
[http://dx.doi.org/10.1128/JVI.79.18.12058-12064.2005] [PMID: 16140781]

[143] Seo SH, Hoffmann E, Webster RG. The NS1 gene of H5N1 influenza viruses circumvents the host anti-viral cytokine responses. Virus Res 2004; 103(1-2): 107-13.
[http://dx.doi.org/10.1016/j.virusres.2004.02.022] [PMID: 15163498]

CHAPTER 3

Influenza Virus Ecology and Evolution

Abstract: Influenza viruses undergo evolution by the shift (influenza A viruses) and the drift of envelope glycoproteins. Since predicting the direction of the antigenic evolution of influenza viruses is impossible, it is necessary to monitor wild viruses in nature. An important objective is to control the circulation of influenza viruses in natural reservoirs, interspecies transmission, especially the acquisition by avian influenza viruses and their ability to infect mammals, and detection of cases of stable transmissibility of such strains. In addition to antigenic evolution, it is necessary to identify and track the patterns of evolution of other biological properties of influenza viruses that might promote the evasion of viruses from the factors of nonspecific defense of the body and aggravate the epidemic process.

Keywords: Avian influenza, Evolutionary variability, Human influenza, Influenza virus, Nature reservoirs, Swine influenza.

INTRODUCTION

The biosphere of the Earth is a common ecosystem, where all living organisms have a mutual influence on each other. The coexistence of people and various animal species in the same space might have not only a positive mutual influence. Between animals and humans quite often occurs the exchange of pathogens. Influenza viruses are among the most common causative agents of human and animal diseases.

INFLUENZA VIRUS RESERVOIRS

Influenza A viruses pose a serious threat to human health and many species of warm-blooded animals (Fig. **1**) [1, 2].

Wild migratory waterfowl and shorebirds are a natural reservoir of influenza A viruses. Among them are common viruses, belonging to 16 serotypes of HA (H1–H16) and 9 serotypes of NA (N1–N9) [3 - 6]. The existence of a large number of different combinations of HA-NA without their genetic linkage to other segments of the genome indicates that mixed infections and reassortment of influenza viruses in wild birds is an ordinary phenomenon. For wild waterfowl,

Irina Kiseleva and Natalie Larionova

influenza viruses of their reservoir are most often low-virulent (LPAI, low pathogenic avian influenza). They are transmitted by the fecal-oral route through contaminated water and infect the epithelial cells of the lower intestine of birds with the development of asymptomatic or mild infection [1, 7].

The bird reservoir plays a key role in the origin of new influenza viruses. Wild bird viruses can cross the interspecies barrier and cause transient zoonotic infections in domestic and land birds and mammals. In rare cases, viruses adapt to new hosts and form their own virus lines (Fig. **1**) [8, 9].

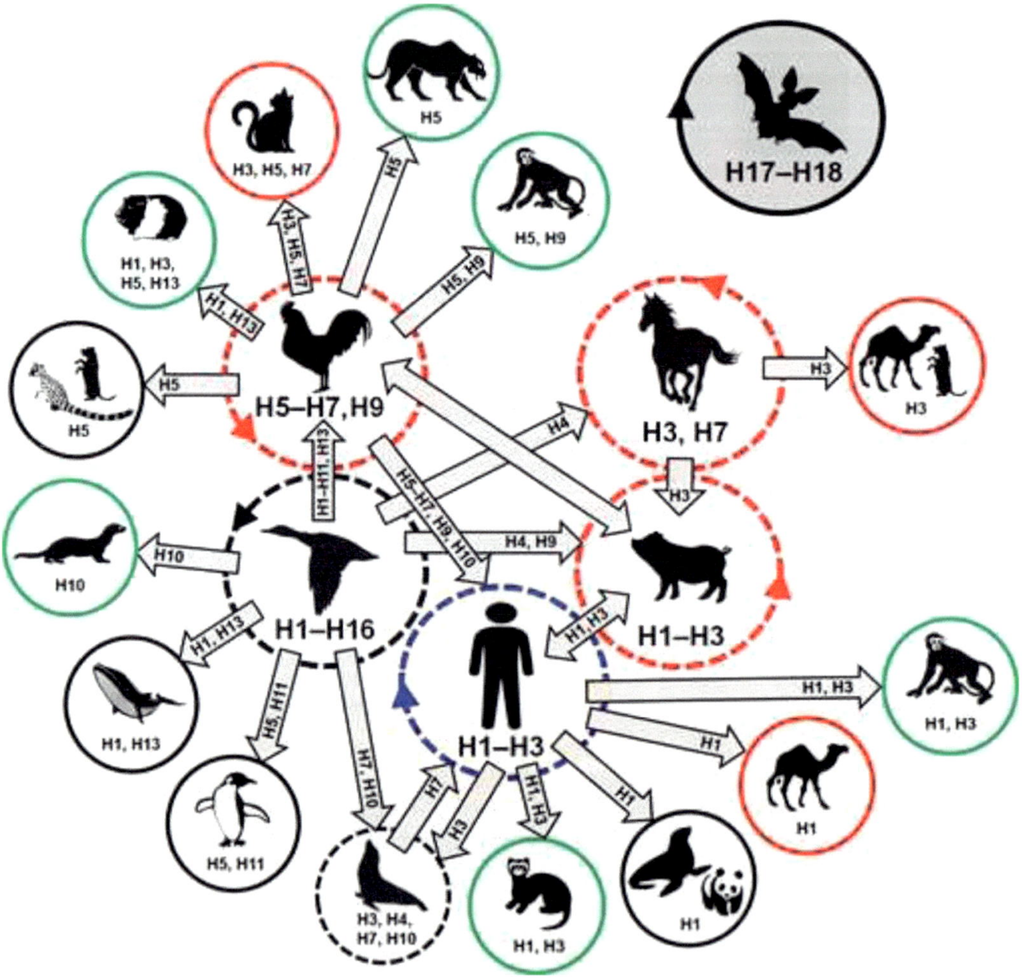

Fig. (1). Low pathogenic avian influenza virus reservoirs and events of their interspecific transmission. The line patterns: dashed line – interspecies transmission described (opened reservoir); solid line – interspecies transmission is not known (closed reservoir). The color patterns: red – poultry, domestic animals; blue – humans; black – wild nature; green – captive animals. The diagram is based on the data presented in a few studies [9 - 34].

Environmental studies and phylogenetic analysis of genome sequences of different serotypes of influenza A viruses isolated from various hosts in different regions confirm the hypothesis that all mammalian influenza viruses originate from the reservoir of avian influenza viruses. The genetic relationship between human and swine influenza viruses also indicates their evolution from a common avian predecessor [35, 36].

Transmissibility of viruses from migratory waterfowl to other species is usually a consequence of mutations and reassortments [36, 37]. In exceptional cases, viruses of low virulence for wild birds, with new mutations can acquire high virulence (HPAI, highly pathogenic avian influenza) for other species of wild and domestic birds and mammals. Outbreaks of HPAI virus disease result in massive deaths in chickens and huge losses in the poultry industry [38].

HPAI H5N1 viruses first declared themselves in 1997 with the mass death of wild and domestic birds [15]. Now, they become a source of epizootics in 77 countries of three continents [17, 38, 39]. This has resulted in tens of millions of avian deaths, infections, and mortality in several mammalian species, including humans [40]. From the vast gene pool of avian influenza viruses, strains stably adapted to mammals have so far formed only four serotypes: H1, H2, and H3 for humans and pigs, H7 and H3 for horses, H3 for dogs [22, 23].

Recent data indicate that the list of animals susceptible to influenza viruses is much wider, and the pathways of interspecies transmission are more complicated than previously thought. Nonhuman primates were found to be naturally infected with seasonal human influenza viruses [19]. Cases of infection of domestic cats, tigers, leopards with HPAI viruses H5N1 were detected [20, 21, 41, 42]. Precedents of infection of minks on farms with influenza viruses of birds, pigs, humans, and cases of interstitial pneumonia, provoked in them by the bird H10N4 virus, have been noted [26, 28]. Unique mutations in the hemagglutinin molecule led to the evolution of the equine H3N8 virus into the canine influenza virus, with the development of severe respiratory disease in the new host [25]. The own line of avian influenza viruses H11N2 has been identified in chinstrap penguins [27].

Pigs play a special role in the emergence of new influenza viruses. Swine influenza viruses are the most common cause of asymptomatic respiratory tract infections in their hosts. Influenza viruses of birds, lower mammals, and humans are periodically transmitted to pigs and successfully reproduced in them. The susceptibility of pigs to influenza viruses of different hosts is due to the presence in the cells of their respiratory tract of both types of receptors: Siaα-2,6Gal and Siaα-2,3Gal. Pigs are figuratively called a "mixing vessel" since the circulation of viruses from different hosts in their population is accompanied by the exchange of genes, mutations, or a complex of these processes [36, 37, 43, 44]. The result of such events may be the emergence of new pathogens, in particular human pathogens, capable of causing major epidemics and global pandemics [45]. The asymptomatic circulation of swine influenza viruses also requires close attention. For example, genetic analysis of viruses isolated from pigs in Hong Kong showed that the precursor to the 2009 pandemic H1N1 virus circulated in pigs 9-17 years before the pandemic [46]. Periodically, swine influenza viruses have become implicated in sporadic human diseases through close contact with pigs. Most often, these diseases did not differ in severity, and transmission from person to person was limited. However, serious illnesses are possible. Significantly, humans can transmit even more viruses to pigs than pigs to humans. In particular, it was shown that the A (H1N1)pdm09 virus circulates in pigs in West Africa after the 2009 year pandemic [47]. Reassortment of human and porcine viruses can lead to the emergence of pathogens that pose a serious threat to both hosts.

Focusing on significantly reducing the possibility of human and animal transmission of influenza viruses to pigs and monitoring the reservoirs of influenza viruses in pigs and wild birds will help determine which of the many influenza viruses can potentially be transmitted to humans and cause a pandemic [43, 48, 49].

Unlike influenza A viruses that are widespread in nature, humans are the primary host and reservoir of influenza B and C viruses [50]. Nevertheless, the facts demonstrate that influenza B and C viruses are capable of transmission from humans to other biological species [30, 51, 52]. There is evidence of the circulation of the influenza C virus among pigs in China [53]. Serological studies in Japan and Great Britain confirm that 9.9% to 19% of pigs have antibodies against the influenza C virus [54, 55]. Cases of reverse zoonotic transmission of A and B influenza viruses to seals after outbreaks in humans have been established [16, 56]. Since 1995, 0.5 to 2% of seals have been infected with the influenza B virus circulating in humans 4-5 years ago [57]. These finds support the notion that seals could serve as a reservoir of human influenza B viruses [16, 56, 57]. Antibodies for influenza A (H1, H3, H5) and B viruses have been detected in guinea pigs kept in farms in Ecuador [30]. Whether this is the result of the adaptation of influenza B viruses to a new host or guinea pigs represent a transitory reservoir of human influenza viruses is not clear yet.

In addition to expanding knowledge about the ecology of the pathogen, in recent years, detection of previously unknown influenza viruses in new natural reservoirs has become a discovery. New hosts of influenza A viruses were discovered in Guatemala. These are two species of bats; influenza A viruses of previously unknown serotypes HA and NA (H17N10 and H18N11) were isolated from their rectal swabs [32 - 34, 58]. Bat viruses have not been found in other mammalian species. However, they have the potential to become a real threat to humans due to the similarity of the replicative complexes of bats and other influenza A viruses, which may be a favorable factor for the formation of interspecies reassortants and interspecies transmissibility [59].

A recent discovery was the isolation of respiratory pathogens in cattle that are distantly related to the influenza C virus. Pigs, sheep, goats, and camels are also susceptible to the virus. These viruses are combined into a new genus Influenzavirus D [60, 61].

There is no doubt that bats as a huge reservoir of natural viruses, whose population is approximately 20% of all mammals, as well as the human neighbor cattle - the reservoir of influenza D viruses - require monitoring [60, 62, 63].

The presented data indicate that the genetic elements of influenza viruses circulate and evolve globally in vast ecosystems that include different hosts and differ in environmental conditions. Influenza A virus gene constellations that are found to be unstable in birds can become stable when switched to a secondary host, to which they can adapt and then circulate independently.

The influenza viruses found in animals and humans have persuaded the researchers these days to give importance to their spread in the biosphere, in order to understand their mechanisms of occurrence, and to find the possible evolutionary pathways of human influenza viruses.

THE ORIGIN AND EVOLUTION OF HUMAN INFLUENZA VIRUSES

Influenza viruses are unique in many of their manifestations: in the ability to infect many host species, evolve, overcome the interspecific barrier, adapt to a new community with the capability to replicate and spread in it, and are unique in the frequency of interspecies transmission events. All these features are realized as a result of the extremely high genetic variability of viruses.

The Genome Evolution

The similarity in the structural features and genome organization of influenza A, B, and C viruses indicate their common ancestry. The mathematical model created to assess the origin and evolution of viral hemagglutinin proves that the weakly virulent influenza C virus, well adapted to the human population, is the oldest, it arose 4000 years earlier than the influenza B virus and 8000 years earlier than the influenza A. The rate of evolution of influenza viruses B and C is lower than that of the younger influenza A. Influenza A virus divergence into serotypes may have occurred about 2000 years ago [64].

The virus, evolutionarily adapted to the host, coexists with it in relative dynamic equilibrium, in which no death or severe morbidity occurs. The good adaptation of influenza A viruses of wild waterfowl to their natural reservoir is evidenced by the minimum amino acid variability of proteins and low virulence [40]. Due to the short life cycle of birds and the low pathogenicity of avian viruses, they are not subjected to antigenic variability. However, the interspecies transmission of influenza A viruses is accompanied by antigenic and genetic evolution. At the same time, non-pathogenic avian viruses, due to the acquisition of adaptive mutations or reassortment with other influenza viruses, can become highly virulent for the new host. Outbreaks of disease in domestic and wildland birds, mammals, and humans are the result of such events. It has been described, in particular, for viruses of the H5 and H7 serotypes [6]. Numerous amino acid substitutions in all proteins of mammalian and land bird viruses confirm that the optimal adaptation of influenza A viruses to these species did not occur [65, 66].

Comparative analysis of a large collection of genome-wide sequences of human influenza A viruses reveals multiple point mutations and deletions in genes encoding coat proteins and internal proteins, as well as replacement of genes entirely due to reassortment [67]. The constant evolution of human influenza pathogens is associated with changing relationships with the host, which are influenced by the genetics of the virus, cell, environmental and social factors. Influenza viruses in an infected organism collide with the innate and adaptive immune press, which activates the renewal of viral antigenic determinants. Changes can also occur with the use of chemotherapy and vaccines.

The two main reasons for the genetic diversity contributing to the rapid evolution of influenza viruses are (**i**) mutations occurring with a high frequency in the genome, and (**ii**) genetic reassortment of genome segments of different influenza viruses.

(**i**). The uncontrolled accumulation of mutations during replication is due to frequent errors in the viral RNA polymerase, since the influenza virus, as a minus-stranded RNA-containing virus, is encoded by a complex of RNA-dependent RNA polymerases, which lacks the function of proofreading and editing. The number of erroneous insertions at nucleotide positions is approximately 10^{-4}-10^{-5} per replicating genome, which corresponds to approximately one altered nucleotide in each cycle of replication [68, 69]. Under the conditions of the population immunological press, the ability of the virus to infect decreases, and the harmful mutations are likely to become an acute problem for influenza viruses, which increases the importance of genetic drift. Most of the emerging mutants and reassortants of the virus are less viable than the parental strains. If spontaneous mutations provide the virus with selective advantages, then they become fixed, and mutants become predominant in the viral progeny [70, 71]. The evolution of genes occurs according to the principle of epistatic interaction, when destabilizing and unacceptable but having an adaptive value, mutations are supported by additional compensatory mutations [72].

(ii). The fragmentation of virion RNA, where each discrete fragment is a gene, promotes the reassortment of genome segments when a cell is infected with two or more strains [73]. The segmented genome provides the virus with evolutionary advantages, which consist in reconstructing an error-free genome due to the ability to select fully copied segments from those copied with errors (selection can occur during assembly of a viral particle or as a result of multiple reactivations) and in the acquisition of favorable alleles. The frequency of reassortment events depends on the dose of infecting viruses, the time of infection, and the compatibility of genome fragments [74]. From 46% to 86% of viral offspring are reassortants when guinea pigs are infected with 10^6PFU of two viruses [75]. The exchange of vRNA segments, in theory, can lead to the formation of $2^8 + 2$ variants of viral progeny and is limited only by the functional incompatibility of genetic components, which in some cases occurs during co-infection with viruses from different hosts [76 - 78]. The genome segments of the influenza A virus serotypes that circulate in humans are fully compatible. The genomes of influenza B Victoria and Yamagata evolutionary lines are compatible too. Therefore, reassortant influenza viruses are frequent participants in the epidemic process [79, 80].

Reassortment is a fundamental factor in the evolution of influenza viruses, playing a key role in the formation of epidemically significant strains and in increasing the diversity of circulating influenza viruses. The segmented genome plays a significant role both in the evolution of influenza viruses within species and in facilitating the transmission to a new host [81, 82].

Antigenic Evolution of Influenza Viruses: Shift and Drift

Antigenically updated viruses acquire the potential to spread in susceptible populations as pathogens of epidemics and pandemics. Two processes determine the antigenic evolution of influenza viruses - shift and drift.

Antigenic shift is the appearance of a completely new antigenic variant of the influenza virus which is unfamiliar to humans, or a return in the circulation of a virus with unknown to the immune system of the next generation of people serotype HA [46, 83, 84]. The introduction to humans of new zoonotic origin HA is due to two mechanisms: reassortment of human and animal viruses and direct interspecies transmission. Only influenza A viruses, which circulate widely in nature among different biological species, are susceptible to the shift. Close contact of wild and domestic waterfowl, animals, and people, animals' markets facilitate the transmission and reassortment of viruses [7, 49, 85]. The spread of a new strain in an immunologically naive population may result in an epidemic on a planetary scale - pandemic.

In the 20th century, there were two hypotheses for the occurrence of influenza pandemics - cyclic and spiral [85]. According to the cyclical hypothesis, the change in hemagglutinins is limited to three serotypes that caused all known to date pandemics: H1, H2, and H3, thereby implying that the next pandemic will be caused by the reappearance of the virus that caused epidemics many years ago.

The spiral change of hemagglutinins implies that viruses of various HA serotypes circulating in animals can cause a pandemic. The global spread of the new virus in the modern world is facilitated by the acceleration of scientific and technological progress with the development of high-speed vehicles, international contacts, and increased migration.

There is no winner in this dispute yet, even possible that the truth is somewhere in between. Although transmission of animal viruses to humans is regularly registered and some viruses, as HPAIV, are highly pathogenic in humans, so far only viruses of the above-mentioned three serotypes have been the cause of pandemics.

The introduction of a new shift influenza virus into the human population is followed by its long-term antigenic evolution – drift [86, 87]. Both influenza A and B viruses are susceptible to drift. Antigenic drift is caused by the gradual accumulation of point mutations in the antigen-presenting sites HA and NA, leading to the ability of the virus to avoid the neutralizing action of the immune press.

During the drift, HA acquires additional glycosylation sites [88]. Thus, over time, H3N2 viruses acquire 11 potential glycosylation sites [89], while the virus-founder of the Hong Kong influenza pandemic had only two such sites [90]. The Spanish flu H1N1pdm1918 had only one 104-Asn glycosylation site, while viruses of this pandemic cycle were characterized by 4–5 sites in the 1950s [91]. The same single 104N glycosylation site was specific for H1N1pdm09 viruses; later, several variants with changes in HA glycosylation were identified [92]. Glycan chains can shield antigenic centers and receptor binding sites and thus facilitate the evasion of the virus from the action of specific antibodies and nonspecific inhibitors in mucous secretions [90, 93].

Influenza Pandemics

Global epidemics similar in description to influenza have been known for at least five centuries [49]. Over the past 100 years, humanity has faced four influenza pandemics, which occurred at intervals of 10 to 40 years. The causative agents of known pandemics were influenza A viruses H1N1, H2N2, and H3N2 with hemagglutinin and neuraminidase of zoonotic origin. The severity of the pandemics ranged from moderate to catastrophic.

All pandemics of the 20th and 21st centuries have resulted from the transmission of zoonotic viruses to humans. Pandemic influenza viruses arose either as a result of a direct adaptation of the avian virus to humans (such is, apparently, the origin of the H1N1 virus, the causative agent of the "Spanish flu" in 1918 [94, 95], or as a result of reassortment of influenza viruses of humans, birds, pigs with an inheritance of zoonotic HA and NA; this is how the pandemic viruses of 1957 (H2N2), 1968 (H3N2) and 2009 (H1N1pdm) originated [37, 83, 96, 97].

The pandemic of "Spanish flu", A (H1N1)pdm1918. The most unprecedented in severity pandemic occurred in 1918-1919 [37]. It killed 40 to 50 million people and caused a huge number of severe forms of the disease. 30% of the world's population suffered from "Spanish flu" [98].

The "Spanish flu" morbidity picture was characterized by unusually high mortality rates among young adults. At the same time, the morbidity structure was similar to other pandemics - children under the age of 15 were at the highest risk. The "Spanish flu" virus was limited by damage to the respiratory tract [99]. Most of the patients died from bacterial pneumonia, which can be attributed to the lack of antibiotics at that time; however, many also died from viral pneumonia. Even though the virus was not isolated during the pandemic, it was possible to sequence it, using the genetic material of the victims of the pandemic preserved in the permafrost, and recreate it using reverse genetics [94, 100]. It was found that the causative agent of the pandemic was an avian-like H1N1 virus, with amino acid tags of the human influenza virus in several proteins [95]. It is possible that for several years before the transition to humans, pigs served as intermediate hosts, facilitating the adaptation of the avian virus to mammals [101].

The pandemic of "Asian influenza", A (H2N2)pdm1957. The "Asian" influenza virus originated in southern China in February 1957 and caused two waves of the pandemic in 1957 and 1958. The causative agent of this pandemic reassortant virus A/Singapore/1/57 arose as a result of the introduction of avian HA (H2), NA (N2), and PB1 genes into the genome of the human influenza virus circulating after the "Spanish flu" [96, 102]. This virus killed ~ 1.5 million people.

The pandemic of "Hong Kong" influenza, A (H3N2)pdm1968. The A/Hong Kong/1/68 (H3N2) virus displaced the H2N2 viruses from circulation. Firstly, it was isolated in July 1968 and caused a pandemic in the winter periods of 1968-1969 and 1969-1970. The new pandemic strain is a reassortant of human and avian influenza viruses. It originated as the result of the inheritance by the virus of the "Asian flu" of the HA (H3) and PB1 genes from the avian influenza virus [96, 102]. The "Hong Kong" flu killed ~ 1 million people.

The pandemic of "Russian flu", A (H1N1) of 1977. A new influenza A(H1N1) virus outbreak began in China in May 1977. In the winter of 1977-1978, the epidemic spread to the Northern Hemisphere. The causative agent A/USSR/90/77, otherwise called "Russian" influenza, isolated in Khabarovsk, the USSR, was very similar to the H1N1 viruses circulating in 1952. The assumptions that immediately appeared about the incident that led to the accidental leak of the strain from the Chinese laboratory have not yet been refuted [103]. The virus caused moderate morbidity, predominantly among young adults born after 1956, who have no immunity to the influenza virus of the 1950s, so the pathogen cannot be considered fully pandemic [44].

The pandemic of 2009, A (H1N1)pdm09. The first pandemic of the 21st century was again caused by the H1N1 serotype virus, which arose as a result of the reassortment of swine influenza viruses of two lines - the Eurasian and American [83, 104, 105]. The Eurasian virus spread to pigs from birds in 1979. The American swine influenza virus is a triple reassortant; it is the product of a combination of genes from three viruses from different hosts. Its ancestors are the classic swine influenza virus (isolated by R.E. Shope in 1930), the North American avian, and the human H3N2 virus, which reassorted in 1998 in North America. Further reassortment of the American and Eurasian swine influenza viruses led to the emergence of a new causative agent of the pandemic. Virus A (H1N1) pdm09 comprises genes NA and M from the Eurasian virus, PB1 from human virus, PB2 and PA from avian virus, and genes HA, NP, NS from the classic swine virus [83].

The re-emergence of H1N1 serotype pandemic influenza in 2009, antigenically and structurally closely related to the causative agent of the 1918 pandemic, but with significantly lower virulence, was unexpected. The first diseases were reported in Mexico in February-March 2009. In mid-April, a prototype of the pandemic virus, the A/California/07/09 influenza virus strain, was isolated in Southern California, USA. Within a few months, the virus spread widely [106], causing millions of cases and, according to WHO estimates, as of March 19, 2010, resulted in more than 16813 laboratory-confirmed deaths [107]. According to the clarifying data [108 - 110], in the first year of the pandemic, the mortality rate was significantly higher and ranged from 151,700 to 579,000 people. The pandemic is considered moderate.

This is explained by the fact that the pathogen was antigenically relatively close to the H1N1 viruses circulating before 1957, and usually the most vulnerable contingent - people over 65 years of age suffered a relatively mild infection, retaining cross-immunity [111]. Young healthy people 25-45 years old and pregnant women had the worst disease [112]. Respiratory infection was complicated by primary viral (in adults) or secondary bacterial (in children) bronchopneumonia and the development of acute distress syndrome. Compared to seasonal H1N1 viruses, H1N1pdm09 is also a more potent inducer of anti-inflammatory mediators. The virus continues to circulate to the present, becoming a seasonal influenza pathogen.

In the 20th century, pandemic experience, the emergence of a new pandemic virus marks the mysterious disappearance of its predecessor, which drifts in seasonal epidemics. Taking the 2009 H1N1 pandemic as an example, a hypothesis was put forward, the essence of which is that the mechanism leading to the elimination of the previous pathogen is an increase in the production of antibodies to the conservative stem of HA [113 - 115]. The membrane-proximal stalk domain of HA, in contrast to the variable head, is conservative among hemagglutinins of the 1st group, among the HA of the 2nd group, and the HA of influenza B viruses. In people who have had seasonal flu, antibodies are produced mainly to the immunodominant region of the globular head of HA, and only a small amount of B-cell immune response specific to the subimmunodominant domain of the stem is formed. The 2009 H1N1 pandemic virus diverged 30% with the preceding seasonal H1N1 virus in the immunodominant region of the HA head. As a result of the 2009 pandemic, the immune response to the region of the conserved stem in both viruses was boosted. This was enough to squeeze out from the population the precursor of a new pandemic and the seasonal H1N1 virus. Probably the same scenario was followed by the replacement of the H1N1 viruses of the previous pandemic cycle by the pandemic strain H2N2 in 1957. Both pathogens have a group 1 HA with different heads, but a conserved stem domain. The disappearance of human strains H2N2 in 1968 after the appearance of the H3N2 virus, which belongs to the 2nd group, maybe due to a similar mechanism with a difference in the antibodies that were probably boosted to epitopes of the more conserved NA [115, 116].

An exception was the co-circulation of H3N2 and H1N1 viruses after 1977, when the H1N1 viruses returned to the community were unable to supplant the H3N2 viruses. Viruses of both serotypes are still circulating among humans. The co-circulation of these viruses led to the spread of H1N2 reassortants in 2001, but the reassortant viruses did not gain an epidemic advantage and subsequently disappeared. Only in 2009 were the H1N1 serotype viruses, which had been drifting since 1977, were replaced by a new, distant variant of the same serotype; H1N1pdm09 [49].

Influenza Epidemics

An influenza pandemic is followed by a period of evolution of its causative agent. Since 1977, seasonal morbidity has been associated with influenza A viruses of the H1N1, H3N2 serotypes, and influenza B viruses. The antigenic variability of influenza viruses poses a serious health threat: the total mortality over the past 100 years from annual influenza epidemics has exceeded even the mortality rates of the three pandemics combined [84, 117].

The most severe morbidity is caused by the H3N2 viruses, although H1N1 viruses and two antigenically divergent influenza virus lines, B/Victoria/2/87 (Victoria) and B/Yamagata/16/88, (Yamagata), often cause severe infections too [118]. Since 2009, the most serious diseases have been associated with the H1N1pdm09 virus, which has become a seasonal pathogen. However, seasonal H3N2 viruses resumed their activity in 2011-2012.

A model of the evolution of influenza pathogens in space and time has been studied based on antigenic and genetic analysis of hemagglutinins of tens of thousands of viruses isolated over many years on six continents [119].

These studies show that there is a continuous circulation of H3N2 viruses in the regions of East and Southeast Asia. Through local epidemics in these geographic areas, which often overlap in time, a common regional epidemic network is created. From China, Southeast Asia, and India, H3N2 influenza viruses spread epidemically to temperate regions every year. Usually, the strains-founders of epidemics first spread in Oceania, North America, and Europe, and later in South America. Outside East, Southeast, and South Asia, the circulation of H3N2 viruses no longer contributes so significantly to long-term viral evolution [120, 121].

The evolution of the influenza A virus genome is characterized by frequent reassortment events and periodic selection, while serotypes H3N2 and H1N1 have different evolutionary dynamics [119].

The global circulation patterns of H1N1 viruses (causative agents of seasonal influenza before 2009), as well as B viruses of the Victoria and Yamagata lines, differ significantly from H3N2 viruses, as evidenced by the analysis of 9604 hemagglutinin sequences of seasonal human influenza viruses isolated from 2000 to 2012. Genetic variants of H3N2 viruses in the same geographical region did not persist between epidemics, but always spread from East and South-East Asia, while genetic variants of A (H1N1) and B viruses persisted locally for several seasons, and in the spread of new variants of these viruses East and Southeast Asia played a limited role. The distribution of H1N1 viruses by geographic region was particularly noticeable in 2004-2005 when three socially circulating genetic lineages spread. Each lineage acquired mutations in HA, which led to antigenic evolution from a phenotype similar to A/New Caledonia/20/1999 to A/Solomon Islands/3/2006. These lines circulated in Southeast Asia (line 1), China (line 2), and India (line 3).

Eventually, the Indian lineage spread throughout the world and circulated until the H1N1pdm09 viruses appeared. Virus lines B Victoria and Yamagata often circulated for several years outside East and Southeast Asia, with no evidence of their penetration from East and Southeast Asia. An illustrative example is the emergence of the North American "Victorian" viruses of the 2006-2007 season directly from the North American viruses of 2005-2006, as well as the origin of the North American "Yamagata" viruses of 2001-2002 directly from the North American viruses of the 2000-2001 season. The lineages of viruses circulating within East and Southeast Asia were characterized by a separate distribution exclusively in this region for more than 1 year. A long period of independent local circulation in the East and Southeast Asia was characteristic of the "Victorian" viruses. There, for more than 5 years, without spreading to other regions, two lineages circulated, which eventually led to the co-circulation of three separate antigenic variants of "Victorian" viruses in different parts of the world during 2007-2008 season.

Phylogenetic analysis reveals a higher rate of nucleotide mutations and amino acid substitutions in H3N2 and H1N1 viruses compared to influenza B viruses. H3N2 viruses are characterized by rapid linear progressive evolution. The antigenic drift of H1N1 viruses is weaker; it is expressed by various branches from the common phylogenetic trunk. Influenza B viruses show greater genealogical diversity [119].

There is a correlation between the antigenic drift rate of viruses and the rate of their global spread. Although the H3N2 and H1N1 viruses have the same rate of genetic evolution, the H1N1 and B antigenic viruses evolved more slowly than the H3N2 viruses. Accordingly, H3N2 viruses are more likely to move between regions than H1N1 and B. The slower global movement of influenza H1N1 and B viruses results in a smaller population of susceptible individuals, a decrease in the age of those infected, and smaller, less frequent epidemics than H3N2 viruses. A consequence of the rapid antigenic drift is a higher frequency of infections, an aging contingent of sick people, but smaller differences in the linear evolution of viruses.

The inverse relationship between the rate of evolution and the genealogical diversity of viruses is due to the fact that the increasing frequency of mutations correlates with antigenic drift, which contributes to the growth of adaptive evolution, leading to a decrease in genetic diversity, correlates with antigenic drift, this promotes the growth of adaptive evolution, leading to a decrease in the genetic diversity of HA [119].

Technocratic means of mobility contribute to the rapid global spread of influenza viruses in the modern world [122].

The epidemic activity of modern influenza B viruses has increased significantly. They are actively evolving and are participants in annual epidemics. A feature of the evolution of influenza B viruses is the systematic appearance of deletions and insertions in the genome [123]. Since the late 1970s, there have been an alternating circulation of two antigenically distinct lines - B/Victoria/1/87- and B/Yamagata/16/88-like viruses [89].

From 1990 to 2001, only Yamagata strains circulated in Europe, America, Africa, and Australia. "Victorian" strains were not widespread and circulated limitedly in the countries of Southeast Asia. In recent years, viruses of the Victoria and Yamagata lineages have been circulating alternately and simultaneously.

Understanding the Spatio-temporal patterns of the emergence, global migration, and circulation of new seasonal variants of human influenza viruses is essential for preventing their spread, monitoring circulating viruses, competently assessing the epidemiological situation, and minimizing the discrepancy between vaccines used for prevention and actual pathogens of epidemics.

Influenza Viruses with Pandemic Potential

The risk of a pandemic is high if a new virus is capable of causing human disease, has antigenic properties, to which the population is immunologically naive, and it can spread steadily between people [124].

Theoretically, all serotypes of influenza viruses circulating among animals, in particular birds and pigs, as possessing antigens unfamiliar to humans, can carry a pandemic threat. It is also likely that a new pandemic can be provoked by viruses of the serotypes H1N1, H2N2, and H3N2 that previously circulated in humans, if a significant layer of people has lost their immunity. Among the causative agents of previous influenza pandemics, viruses of the HA H2 serotype, which last caused an epidemic almost 50 years ago, may pose a threat. Porcine/avian viruses of the H2N3 serotype continue to circulate in pigs, they have α-2,6 receptor specificity [14], so their introduction into the human population seems quite likely.

Most influenza viruses of zoonotic origin that sporadically infect humans through close contact with animals cannot effectively spread by aerosol between humans [125], however, the adaptation of zoonotic viruses to humans through mutation or reassortment with human influenza viruses can lead to respiratory transmission strains. Among zoonotic pathogens, avian viruses of serotypes H5, H7, H9 pose the greatest threat.

Avian influenza viruses. HPAIV H5N1 first caused fatal human diseases in 1997 in Hong Kong. Out of 18 infected cases, 6 died [126, 127]. The genetic characteristics of H5N1 viruses isolated in 1997 from humans and poultry suggest that they are reassortants that acquired the genes from avian H9N2 or H6N1 viruses, and N1 NA from avian H6N1 virus [128]. The outbreak was brought under control as a result of the elimination of a huge number of birds. After a period of sporadic localized infections, a new outbreak began in 2003. Since then, HPAIV H5N1 has evolved rapidly. According to the antigenic characteristics of H5 HA viruses circulating since 2008 attributed to 10 clades (0-9) and numerous subclades [129]. They often entered reassortment [128, 130], spread throughout Europe and Africa, and began to infect the poultry in Southeast Asia [131].

HPAI H5N1 viruses have several alarming features [37]:

(i) cause lethality not only in poultry but also in wild waterfowl, which are a natural reservoir of influenza A viruses;

(ii) capable of replication in mice with the development of a lethal infection without preliminary adaptation;

(iii) infect a number of mammalian species with a fatal outcome, registered facts of non-lethal infections of pigs;

(iv) their pathogenicity for ferrets has increased over the years, which indicates the acquisition of mutations that are virulent for mammals;

(v) of greatest concern is cases of HPAIV transmissibility to humans, resulting in severe lower respiratory tract infections with pneumonia and high mortality.

From 2003 to 10.07.2020, HPAI H5N1 viruses killed 455 people out of 861 cases reported (53%) [132]. Several cases of intrafamilial spread of the infection have been described, but there has been no sustained human-to-human transmission of the virus [133]. Thus, H5N1 viruses are characterized by a high mortality rate, but, at least so far, they are not very efficient in spreading among humans.

Since 2014 human morbidity and a high lethality have also been caused by HPAI viruses of the H5N6 serotype, which are reassortants of avian H5N1 and H6N6 viruses [134, 135].

And since 2010, HPAIV H5N8 has been circulating among wild birds in Asia. In 2014-2015 this virus spread across Europe and North America. No cases of human infection have been reported, and no marks of the pathogenicity of this virus for humans have been found [136].

The apathogenic to birds H7N9 virus has proven to be highly virulent to humans. Among people, that were infected with viruses of the H7N9 serotype since its spread in 2013, a fatality rate of 39% was recorded (616 deaths among 1568 infected patients) [137]. All infections are the result of direct human contact with poultry. Evolutionary analysis has shown that the H7N9 virus isolated from humans is a reassortant derived from avian viruses H9N2, H7N3, and H11N9. The internal genes of the H7N9 and H5N1viruses belong to the same genetic line [138]. The virus carries a potential threat to humans also because, due to its low pathogenicity, it is not tracked in birds, but at the same time, it can be transmitted to humans [139].

It is alarming that the amino acid sequence of the proteins of the avian H7N9 virus contains signs that indicate the possibility of transmissibility of the virus between humans: in the position of the PB2-627 protein there is Lys, which is inherent in human viruses. The PA protein also contains labels characteristic of human viruses: PA-100A, PA-356R, and PA-409N [10]. It was also shown that some H7N9 isolates from humans were capable of airborne transmission between ferrets [140]. Since 2013, the virus has also infected poultry, spreading by contact, and through an infected environment. And this is also alarming since poultry can become a "bridge" for the adaptation of the virus to humans. The increase in symptomatic manifestations during the two waves of incidence in 2013 led to the assumption of the possibility of limited transmissibility of the virus between people [141]. However, there is still no sustained human-to-human transmission of the virus, and no human cases of H7N9 have been reported outside of China. To date, viruses of serotype H7, like H5, are characterized by high lethality for humans but do not differ in their spreading efficiency.

Viruses of the H9N2 serotype circulating among poultry have been causing human cases in China and Egypt since 1998 [142]. By themselves, H9N2 viruses cause relatively mild infections in people in contact with poultry, but they donate genes for the H5N1, H7N9, and H10N8 serotypes infecting humans. The special role of H9N2 viruses in interactions between animals and humans may be due to a wide range of hosts, including both birds and mammals, and their extensive participation in reassortment [143].

To date, there is information on cases of human infection with avian influenza serotypes H5N1, H5N6, H5N8, H6N1, H7N3, H7N4, H7N7, H7N9, H9N2, H10N8. In some cases, infections have been fatal [1, 3, 22, 144 - 152].

Swine influenza viruses periodically spread quite effectively among humans, but they cause a limited number of fatal infections [36, 153]. The antigenic properties of H3 influenza viruses in humans and pigs indicate that population immunity is not able to prevent the introduction of influenza viruses from humans to pigs and *vice versa* [154].

In 2011, the reassortant H3N2v swine influenza virus, which also contained avian, human viruses genes, and M-gene of the A (H1N1)pdm09 virus, caused human disease in the United States. In 2012, at the height of the outbreak, 309 cases were reported, with one death. The presence of the M gene of the pandemic influenza virus and the high vulnerability of the victims (they all belonged to risk groups) probably facilitated the possibility of human infection [155]. Diseases with mild symptoms caused by H1N1v, H1N2v, H3N2v swine influenza viruses are recorded in humans, mainly in the USA, Brazil and Vietnam [142].

The above facts indicate that zoonotic influenza viruses of various serotypes, as pathogens antigenically unfamiliar to the human immune system, pose a huge potential threat to human health. However, to realize this threat, in addition to antigenic novelty, they must acquire the ability to infect the cells of the respiratory tract and spread steadily from person to person.

EVOLUTIONARY VARIABILITY OF INFLUENZA VIRUS PROPERTIES AS A FACTOR OF HOST-PATHOGEN INTERACTIONS

Evolution of Influenza Viruses Based on Their Sensitivity to Replication Temperature

The driving force behind the evolution of epidemic influenza viruses is the ability to resist the neutralizing effect of not only specific immunity factors, but also nonspecific virus suppressing effects. Mutations aimed at reproduction (self-preservation) can lead to the emergence of virions with altered biological properties, which can be expressed by new manifestations of diseases caused by viruses. The study of influenza A and B viruses circulating in the community showed that there is a fairly extensive genotypic and phenotypic variability among them [156]. In particular, variants were found that differ in virulence.

An important but poorly studied characteristic of influenza viruses associated with the knowledge of the patterns of their evolutionary variability and pathogenicity is the ability to reproduce at temperatures outside the upper limit of optimal values. The experiment revealed that the maximum permissible temperature limiting replication of influenza A viruses is 40°C and influenza B viruses - 38°C.

Observations from the discovery of influenza A and B viruses to the early 1970s of the 20th century have shown that all influenza viruses, including the causative agents of the 1930s-40s H1N1 epidemics, H2N2, and the earliest of the H3N2 pandemic viruses, are capable of reproducing at elevated temperatures [157 - 160]. This led to the conclusion that human influenza viruses are always resistant to high incubation temperatures, which determines their virulence and allows them to withstand the organism's defense factors [161, 162].

Temperature-sensitive (*ts*) mutants of the influenza virus are primarily of interest in the development of attenuated reassortant strains for live influenza vaccines. Laboratory-derived *ts* viruses are always attenuated for humans, and laboratory adaptation to low temperatures has become a successful method for creating vaccine strains [163, 164].

For the first time, the discovery of a natural temperature-sensitive influenza virus was noted in 1973, when a *ts* clone was isolated from a population of a temperature-resistant (*non-ts*) epidemic virus A (H3N2) in developing chicken embryos (CE) without special adaptation [161].

Then, the facts of a wide circulation of sensitive to 40°C A (H3N2) strains were discovered during the 1975-1976 epidemic [165, 166]. In the 1977-1978 epidemic period, 17 out of 26 H1N1 isolates and 2 out of 11 H3N2 isolates had the *ts* phenotype. Even viruses isolated in one city and even from one individual varied significantly in temperature sensitivity. The H1N1 virus with a reduced ability to replicate at 38°C in MDCK cells was partially attenuated for humans. The *ts* mutant, with a limiting reproduction temperature of 37°C, was even more attenuated in the volunteer test. It was concluded that the appearance of viruses with the *ts* phenotype, along with *non-ts*, correlates with the observation that viruses of varying virulence naturally circulate in the community [165].

Research by Chu *et al.* [161] on seronegative volunteers confirmed that *ts* A (H1N1) viruses with upper limiting temperature less than 38°C caused mild infections, while the isolate with the maximum permissible reproduction temperature of 39°C was significantly more reactogenic. The authors also concluded that influenza viruses differing in temperature sensitivity and, probably, of different virulence are widespread in the population.

The temperature sensitivity of laboratory strains is determined by the functioning of the polymerase complex genes [167 - 169]. The molecular mechanisms underlying the manifestation of temperature sensitivity by circulating seasonal human influenza viruses have not been studied.

Such studies were carried out on avian viruses that function at high temperatures (41.5°C) of the host intestine. Reconstruction of functional ribonucleoprotein *in vivo* made it possible to find out that polymerase complexes of avian viruses are sensitive to temperature drops to 33-37°C, which is characteristic of mammalian cells. The decrease in the ability of the polymerase complex of avian viruses to replicate at 33°C is determined by the amino acid residue 627 of PB2 [170]. Similar studies of recombinants between avian and human viruses obtained by genetic engineering methods showed that the RNP complex of the H5N1 virus, into which the PB2 protein from the human H1N1pdm09 or H3N2 viruses was incorporated, increased the polymerase activity in human cells. A single amino acid substitution 627K in the PB2 protein of the H5N1 virus influenced the manifestation of the temperature-dependent mechanism in the polymerase activity. The substitution PB2 E158Q was also involved in the temperature-dependent mechanism of regulation of transcription and replication.

In other experiments, determinants that prevent the replication of human viruses when the temperature rises to 41°C were investigated [171]. The authors found that at non-permissive temperatures, the block of viral reproduction occurs in the late stages of the replication cycle. And although all virus-specific components (vRNA, cRNA, proteins) are produced, maturation of infectious particles does not occur. In the case of A/FM/1/47 (H1N1) and A/Eq/Miami/63 (H3N8) strains, budding did not occur. In the case of A/PR/8/34 (H1N1) virus, hemagglutinating non-infectious filamentous particles up to 6 μm in length were produced, in which the content of NP and M proteins was significantly reduced.

Among influenza A (H7N1) viruses pathogenic for chickens, one was identified to be sensitive to 41.5°C. It is characterized by reduced pathogenicity. Analysis of reassortants between this strain and the highly pathogenic isolates showed that the decrease in pathogenicity is associated with a mutation of the PB2 gene at position 1561, which led to the amino acid substitution I512L [172].

In all the studies described, the sensitivity of viruses to the temperature of reproduction is connected with mutations in the protein PB2.

Cyclical Nature of Evolutionary Variability of Virus Properties

Further observations found that the proportion of ts viruses steadily increased over the years. For instance, the ts isolates accounted for 8.3% of all investigated A

(H1N1) viruses in 1949-1957, while they made up to 76.5% in 1977-1978 [161]. A similar pattern was observed for H3N2 viruses. 34.8% of strains isolated in 1968-1978 displayed the ts phenotype. In 1979-1980 the percentage of ts H3N2 strains reached 82.4% [161].

The pandemic A/Hong Kong/1/68 (H3N2) influenza virus was characterized by temperature-resistant replication at 40°C. Its drift variants retained the *non-ts* phenotype until 1975. 61% of the investigated isolates of 1976 were temperature-sensitive, like their reference representative A/Moscow/406/76 (H3N2) (Fig. **2**).

The standard of the 1977 epidemic, A/Texas/1/77 (H3N2), was temperature-resistant, but already in 1979, it gave way to the sensitive to high incubation temperatures A/Bangkok/1/79 and A/Bangkok/2/79 viruses. Temperature-sensitive A (H3N2) viruses predominated in circulation until 1984. 73% of the studied representatives of this period were characterized by the *ts* phenotype, including the reference virus A/Philippines/2/82 (H3N2).

During 1985-1986, among the circulating H3N2 strains, heterogeneity was observed. The heterogeneity in the *ts* phenotype corresponded to the antigenic heterogeneity of the H3N2 viruses circulating in this time interval: along with temperature-sensitive viruses, such as the reference A/Mississippi/1/85 and A/USSR/2/85, there were isolates of the previous epidemic cycles that differed in the *ts* phenotype and were closely related A/Texas/1/77, A/Bangkok/1/79 and A/Philippines/2/82 [173].

In 1987-1995, the circulation of temperature-sensitive H3N2 viruses continued, including the reference strains A/Sichuan/2/87, A/Shangdong/9/93, A/Johannesburg/33/94, A/Nanchang/933/95. They were replaced in 1997 by H3N2 strains resistant to elevated temperatures, the circulation of which was tracked until 1999 (representatives - reference viruses A/Sydney/5/97, A/Panama/2007/99, A/Moscow/10/99).

From 2000 to 2004, H3N2 viruses, differing in temperature sensitivity, circulated alternately and simultaneously. Recently, temperature-sensitive H3N2 strains still prevail in circulation (Fig. **2**).

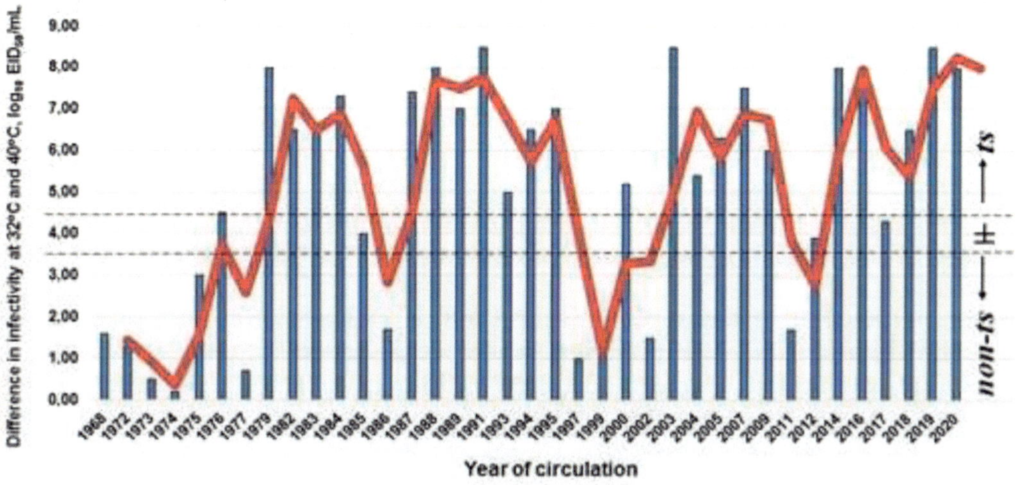

Fig. (2). Evolution of influenza A (H3N2) viruses based on the temperature sensitivity of replication in developing chick embryos. Replication activity of influenza viruses is presented as a mean value of differences in infectivity at 32°C and 40°C, $\log_{10}$ EID_{50}; not less than 5-10 viruses isolated per year were tested. The red trend line shows the change of the *ts* phenotype over time.

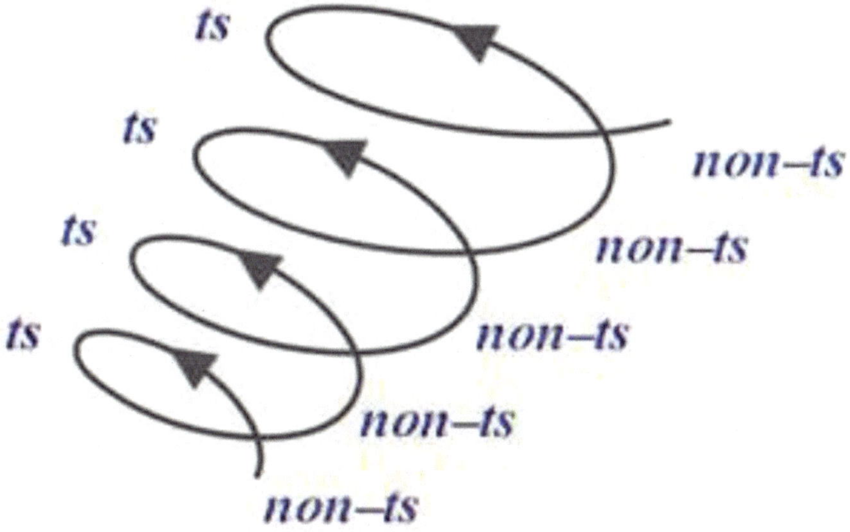

Fig. (3). Scheme of the evolution of influenza viruses based on their sensitivity to the temperature of incubation. Temperature resistant (*non–ts*) influenza viruses wave appear and disappear in the process of evolution.

A similar pattern was also observed among influenza A viruses of other pandemic cycles (H1N1, H2N2) and influenza B viruses [173, 174]. At the beginning of the pandemic cycle in influenza A (H2N2) viruses, early representatives of influenza A (H1N1) and B viruses were invariably characterized by the circulation of temperature-resistant strains. Further evolution led to the alternating circulation of *non-ts* and *ts* viruses. The evolution of influenza viruses in terms of the temperature characteristics of reproduction is represented as a cyclical change in circulating phenotypes: *non-ts* → *ts* → *non-ts* → *ts*, *etc*. Between the next change in the temperature sensitivity of viruses, the circulation of strains characterized by both *non-ts* and *ts* phenotypes, as well as having ±*ts* was observed (Fig. **3**) [173].

Each serotype of influenza A virus (H1N1 and H3N2) and both influenza B virus lines (B/Victoria and B/Yamagata) underwent several wave-like cycles of temperature sensitivity. Even the "Asian" influenza A (H2N2) viruses, despite a short, only eleven-year circulation period, were characterized by an evolutionary transition from the *non-ts* to the *ts* phenotype [173].

Over time, viruses varying in the sensitivity to the replication temperature were finally replaced by temperature-sensitive viruses. It is noteworthy that on the eve of the A (H1N1)pdm09 pandemic, from 2005 to 2009, circulating strains of serotype A (H3N2), as well as influenza B viruses, were characterized by *ts* phenotype. The same trend was noted for the A (H1N1) viruses, among which, since 2006, only *ts* viruses have been isolated. The *ts* viruses circulated until 2009 when they were superseded by rapidly spreading and causing a new pandemic, the high temperature-resistant virus A/California/07/09 (H1N1)pdm.

Pandemic virus A/California/07/09 (H1N1)pdm has opened another cycle of circulation of *non-ts* viruses - it has high reproductive capacity at 40°C (the difference in the infectious titer of the virus at 32° and 40°C is only 1.2 log EID_{50}/mL). Antigenically closely related A (H1N1)pdm09 viruses, having become seasonal pathogens, continue to maintain a temperature-resistant phenotype up to the present.

Temperature-resistant viruses reappeared among the A (H3N2) viruses, starting with the A/Victoria/361/11 virus, and among the influenza B viruses B/Massachusetts/2/12 (B/Yamagata lineage), B/Texas/02/13 (line B/Victoria).

The potential ability of human influenza A viruses to replicate at temperatures elevated to 39-40°C is probably due to the origin of at least two of the three genes of the temperature-dependent polymerase complex from influenza viruses of animals (birds or pigs), which have a higher physiological body temperature (40-42°C). When adapting to humans, the temperature range of virus reproduction is probably gradually reconfigured. Influenza B viruses with anthroponotic tropism have a lower maximum replication temperature (38°C), although they are also subject to cyclical changes.

The epidemic potential of viruses is determined by their novelty for the human immune system. However, with the growth of population immunity to circulating viruses, evolutionary selection is likely to take place in favor of those with a higher virulence potential.

If the *ts* phenotype of circulating influenza viruses indirectly reflects a decrease of their virulence, and together with the growth of the immune layer in society, reduces their epidemic potential, then the fact of stable circulation for several years of temperature-sensitive viruses can be considered as a prerequisite for the appearance of a new antigenic variant - shift or antigenically distant from previously circulating drift viruses. This assumption is supported by the prolonged circulation of temperature-sensitive influenza A (H1N1), A (H3N2), and B viruses before the start of the 2009 pandemic.

It seems reasonable that the evolution of the influenza viruses should assess not only from the position of the novelty of surface antigens but also by taking into account changes in their biological characteristics, such as the temperature sensitivity of replication.

Taking into account the *ts* phenotype of circulating influenza viruses, it is possible to more accurately assess the epidemic potential of newly emerging strains, to formulate in advance a more substantiated forecast of the expected incidence of influenza, and even to predict the possibility of the appearance of an antigenically renewed influenza virus.

CONCLUDING REMARKS

Zoonotic influenza viruses pose a huge potential threat to human health as antigenically unfamiliar pathogens. To implement this threat, in addition to antigenic novelty, newly emerged viruses must acquire the ability to infect the cells of the respiratory tract and be able to sustain transmission from person to person. Reassortment of animal and human influenza viruses can facilitate the adaptation process. Therefore, constant monitoring of influenza viruses circulating in animals and reservoirs of influenza viruses in nature is very important

Analysis of the variability of the biological properties of influenza viruses brings us closer to understanding the possible ways of their evolution, which contributes to the development of rational methods to prevent the spread of viruses and the morbidity they cause.

REFERENCES

[1] Taubenberger JK, Kash JC. Influenza virus evolution, host adaptation, and pandemic formation. Cell

Host Microbe 2010; 7(6): 440-51.
[http://dx.doi.org/10.1016/j.chom.2010.05.009] [PMID: 20542248]

[2] Chambers TM, Dubovi EJ, Donis RO. Equine/Canine/Feline/Seal influenza.Textbook of influenza. 2nd ed. UK: Wiley Blackwell 2013; pp. 203-17.
[http://dx.doi.org/10.1002/9781118636817.ch13]

[3] Fouchier RA, Schneeberger PM, Rozendaal FW, *et al.* Avian influenza A virus (H7N7) associated with human conjunctivitis and a fatal case of acute respiratory distress syndrome. Proc Natl Acad Sci USA 2004; 101(5): 1356-61.
[http://dx.doi.org/10.1073/pnas.0308352100] [PMID: 14745020]

[4] Naguib MM, Verhagen JH, Mostafa A, *et al.* Global patterns of avian influenza A (H7): virus evolution and zoonotic threats. FEMS Microbiol Rev 2019; 43(6): 608-21.
[http://dx.doi.org/10.1093/femsre/fuz019] [PMID: 31381759]

[5] Olsen B, Munster VJ, Wallensten A, Waldenström J, Osterhaus AD, Fouchier RA. Global patterns of influenza a virus in wild birds. Science 2006; 312(5772): 384-8.
[http://dx.doi.org/10.1126/science.1122438] [PMID: 16627734]

[6] Yoon SW, Webby RJ, Webster RG. Evolution and ecology of influenza A viruses. Curr Top Microbiol Immunol 2014; 385: 359-75.
[http://dx.doi.org/10.1007/82_2014_396] [PMID: 24990620]

[7] Webster RG, Peiris M, Chen H, Guan Y. H5N1 outbreaks and enzootic influenza. Emerg Infect Dis 2006; 12(1): 3-8.
[http://dx.doi.org/10.3201/eid1201.051024] [PMID: 16494709]

[8] Parrish CR, Holmes EC, Morens DM, *et al.* Cross-species virus transmission and the emergence of new epidemic diseases. Microbiol Mol Biol Rev 2008; 72(3): 457-70.
[http://dx.doi.org/10.1128/MMBR.00004-08] [PMID: 18772285]

[9] Short KR, Richard M, Verhagen JH, *et al.* One health, multiple challenges: The inter-species transmission of influenza A virus. One Health 2015; 1: 1-13.
[http://dx.doi.org/10.1016/j.onehlt.2015.03.001] [PMID: 26309905]

[10] Chen GW, Lai MM, Wu SC, *et al.* Is avian influenza A (H7N9) virus staggering its way to humans? J Formos Med Assoc 2013; 112: 312-8.
[http://dx.doi.org/10.1016/j.jfma.2013.04.015] [PMID: 23742902]

[11] Amonsin A, Payungporn S, Theamboonlers A, *et al.* Genetic characterization of H5N1 influenza A viruses isolated from zoo tigers in Thailand. Virology 2006; 344(2): 480-91.
[http://dx.doi.org/10.1016/j.virol.2005.08.032] [PMID: 16194557]

[12] Mushtaq MH, Juan H, Jiang P, *et al.* Complete genome analysis of a highly pathogenic H5N1 influenza A virus isolated from a tiger in China. Arch Virol 2008; 153(8): 1569-74.
[http://dx.doi.org/10.1007/s00705-008-0145-3] [PMID: 18592132]

[13] Boyce WM, Mena I, Yochem PK, *et al.* Influenza A(H1N1)pdm09 virus infection in marine mammals in California. Emerg Microbes Infect 2013; 2(6): e40.
[http://dx.doi.org/10.1038/emi.2013.40] [PMID: 26038474]

[14] Ma W, Vincent AL, Gramer MR, *et al.* Identification of H2N3 influenza A viruses from swine in the United States. Proc Natl Acad Sci USA 2007; 104(52): 20949-54.
[http://dx.doi.org/10.1073/pnas.0710286104] [PMID: 18093945]

[15] Neumann G, Chen H, Gao GF, Shu Y, Kawaoka Y. H5N1 influenza viruses: outbreaks and biological properties. Cell Res 2010; 20(1): 51-61.
[http://dx.doi.org/10.1038/cr.2009.124] [PMID: 19884910]

[16] Ohishi K, Ninomiya A, Kida H, *et al.* Serological evidence of transmission of human influenza A and B viruses to Caspian seals (Phoca caspica). Microbiol Immunol 2002; 46(9): 639-44.
[http://dx.doi.org/10.1111/j.1348-0421.2002.tb02746.x] [PMID: 12437032]

[17] Swayne DE. Impact of vaccines and vaccination on global control of avian influenza. Avian Dis 2012; 56(4) (Suppl.): 818-28.
[http://dx.doi.org/10.1637/10183-041012-Review.1] [PMID: 23402099]

[18] Chu DKW, Perera RAPM, Ali A, *et al.* Influenza A virus infections in dromedary camels, Nigeria and Ethiopia, 2015-2017. Emerg Infect Dis 2020; 26(1): 173-6.
[http://dx.doi.org/10.3201/eid2601.191165] [PMID: 31855544]

[19] Karlsson EA, Engel GA, Feeroz MM, *et al.* Influenza virus infection in nonhuman primates. Emerg Infect Dis 2012; 18(10): 1672-5.
[http://dx.doi.org/10.3201/eid1810.120214] [PMID: 23017256]

[20] Keawcharoen J, Oraveerakul K, Kuiken T, *et al.* Avian influenza H5N1 in tigers and leopards. Emerg Infect Dis 2004; 10(12): 2189-91.
[http://dx.doi.org/10.3201/eid1012.040759] [PMID: 15663858]

[21] Seiler BM, Yoon KJ, Andreasen CB, Block SM, Marsden S, Blitvich BJ. Antibodies to influenza A virus (H1 and H3) in companion animals in Iowa, USA. Vet Rec 2010; 167(18): 705-7.
[http://dx.doi.org/10.1136/vr.c5120] [PMID: 21257487]

[22] Taubenberger JK, Morens DM. The once and future pandemic. Public Health Rep 2010; 125(Suppl 3): 16-26.
[http://dx.doi.org/10.1177/00333549101] [PMID: 20568566]

[23] Zambon MC. The pathogenesis of influenza in humans. Rev Med Virol 2001; 11(4): 227-41.
[http://dx.doi.org/10.1002/rmv.319] [PMID: 11479929]

[24] Barriga GP, Boric-Bargetto D, San Martin MC, *et al.* Avian influenza virus H5 strain with North American and Eurasian lineage genes in an antarctic penguin. Emerg Infect Dis 2016; 22(12): 2221-3.
[http://dx.doi.org/10.3201/eid2212.161076] [PMID: 27662612]

[25] Crawford PC, Dubovi EJ, Castleman WL, *et al.* Transmission of equine influenza virus to dogs. Science 2005; 310(5747): 482-5.
[http://dx.doi.org/10.1126/science.1117950] [PMID: 16186182]

[26] Englund L. Studies on influenza viruses H10N4 and H10N7 of avian origin in mink. Vet Microbiol 2000; 74(1-2): 101-7.
[http://dx.doi.org/10.1016/S0378-1135(00)00170-X] [PMID: 10799782]

[27] Hurt AC, Su YCF, Aban M, *et al.* Evidence for the introduction, reassortment, and persistence of diverse influenza A viruses in Antarctica. J Virol 2016; 90(21): 9674-82.
[http://dx.doi.org/10.1128/JVI.01404-16] [PMID: 27535050]

[28] Peng L, Chen C, Kai-yi H, *et al.* Molecular characterization of H9N2 influenza virus isolated from mink and its pathogenesis in mink. Vet Microbiol 2015; 176(1-2): 88-96.
[http://dx.doi.org/10.1016/j.vetmic.2015.01.009] [PMID: 25655813]

[29] Hinshaw VS, Bean WJ, Geraci J, Fiorelli P, Early G, Webster RG. Characterization of two influenza A viruses from a pilot whale. J Virol 1986; 58(2): 655-6.
[http://dx.doi.org/10.1128/jvi.58.2.655-656.1986] [PMID: 3701925]

[30] Leyva-Grado VH, Mubareka S, Krammer F, Cárdenas WB, Palese P. Influenza virus infection in guinea pigs raised as livestock, Ecuador. Emerg Infect Dis 2012; 18(7): 1135-8.
[http://dx.doi.org/10.3201/eid1807.111930] [PMID: 22710350]

[31] Molini U, Aikukutu G, Roux JP, *et al.* Avian influenza H5N8 outbreak in African penguins (Spheniscus demersus), Namibia, 2019. J Wildl Dis 2020; 56(1): 214-8.
[http://dx.doi.org/10.7589/2019-03-067] [PMID: 31483707]

[32] Teeling EC, Springer MS, Madsen O, Bates P, O'Brien SJ, Murphy WJ. A molecular phylogeny for bats illuminates biogeography and the fossil record. Science 2005; 307(5709): 580-4.
[http://dx.doi.org/10.1126/science.1105113] [PMID: 15681385]

[33] Tong S, Li Y, Rivailler P, *et al.* A distinct lineage of influenza A virus from bats. Proc Natl Acad Sci USA 2012; 109(11): 4269-74.
[http://dx.doi.org/10.1073/pnas.1116200109] [PMID: 22371588]

[34] Tong S, Zhu X, Li Y, *et al.* New world bats harbor diverse influenza A viruses. PLoS Pathog 2013; 9(10): e1003657.
[http://dx.doi.org/10.1371/journal.ppat.1003657] [PMID: 24130481]

[35] Gammelin M, Altmüller A, Reinhardt U, *et al.* Phylogenetic analysis of nucleoproteins suggests that human influenza A viruses emerged from a 19th-century avian ancestor. Mol Biol Evol 1990; 7(2): 194-200.
[PMID: 2319943]

[36] Webby R, Richt J. Influenza in swine.Textbook of influenza. 2nd ed. UK: Wiley Blackwell 2013; pp. 190-202.
[http://dx.doi.org/10.1002/9781118636817.ch12]

[37] Neumann G, Noda T, Kawaoka Y. Emergence and pandemic potential of swine-origin H1N1 influenza virus. Nature 2009; 459(7249): 931-9.
[http://dx.doi.org/10.1038/nature08157] [PMID: 19525932]

[38] Avian influenza portal 2020.https://www.oie.int/animal-health-in-the-world/web-portal-on-av ian-influenza/

[39] Li C, Bu Z, Chen H. Avian influenza vaccines against H5N1 'bird flu'. Trends Biotechnol 2014; 32(3): 147-56.
[http://dx.doi.org/10.1016/j.tibtech.2014.01.001] [PMID: 24491922]

[40] Webster RG, Hulse-Post DJ, Sturm-Ramirez KM, *et al.* Changing epidemiology and ecology of highly pathogenic avian H5N1 influenza viruses. Avian Dis 2007; 51(1) (Suppl.): 269-72.
[http://dx.doi.org/10.1637/7641-050206R.1] [PMID: 17494564]

[41] Kuiken T, Rimmelzwaan G, van Riel D, *et al.* Avian H5N1 influenza in cats. Science 2004; 306(5694): 241.
[http://dx.doi.org/10.1126/science.1102287] [PMID: 15345779]

[42] Songserm T, Amonsin A, Jam-on R, *et al.* Avian influenza H5N1 in naturally infected domestic cat. Emerg Infect Dis 2006; 12(4): 681-3.
[http://dx.doi.org/10.3201/eid1204.051396] [PMID: 16704821]

[43] Ma W, Kahn RE, Richt JA. The pig as a mixing vessel for influenza viruses: Human and veterinary implications. J Mol Genet Med 2008; 3(1): 158-66.
[PMID: 19565018]

[44] Wang TT, Palese P. Emergence and evolution of the 1918, 1957, 1968, and 2009 pandemic virus strains.Textbook of influenza. 2nd ed., UK: Wiley Blackwell 2013.
[http://dx.doi.org/10.1002/9781118636817.ch14]

[45] Zhou NN, Senne DA, Landgraf JS, *et al.* Genetic reassortment of avian, swine, and human influenza A viruses in American pigs. J Virol 1999; 73(10): 8851-6.
[http://dx.doi.org/10.1128/JVI.73.10.8851-8856.1999] [PMID: 10482643]

[46] Smith GJ, Vijaykrishna D, Bahl J, *et al.* Origins and evolutionary genomics of the 2009 swine-origin H1N1 influenza A epidemic. Nature 2009; 459(7250): 1122-5.
[http://dx.doi.org/10.1038/nature08182] [PMID: 19516283]

[47] Adeola OA, Olugasa BO, Emikpe BO. Detection of pandemic strain of influenza virus (A/H1N1/pdm09) in pigs, West Africa: implications and considerations for prevention of future influenza pandemics at the source. Infect Ecol Epidemiol 2015; 5: 30227.
[http://dx.doi.org/10.3402/iee.v5.30227] [PMID: 26715380]

[48] Nelson MI, Vincent AL. Reverse zoonosis of influenza to swine: new perspectives on the human-

animal interface. Trends Microbiol 2015; 23(3): 142-53.
[http://dx.doi.org/10.1016/j.tim.2014.12.002] [PMID: 25564096]

[49] Monto AS, Fukuda K. Lessons from influenza pandemics of the last 100 years. Clin Infect Dis 2020; 70(5): 951-7.
[http://dx.doi.org/10.1093/cid/ciz803] [PMID: 31420670]

[50] Jackson D, Elderfield RA, Barclay WS. Molecular studies of influenza B virus in the reverse genetics era. J Gen Virol 2011; 92(Pt 1): 1-17.
[http://dx.doi.org/10.1099/vir.0.026187-0] [PMID: 20926635]

[51] Bodewes R, Kreijtz JH, Rimmelzwaan GF. Yearly influenza vaccinations: a double-edged sword? Lancet Infect Dis 2009; 9(12): 784-8.
[http://dx.doi.org/10.1016/S1473-3099(09)70263-4] [PMID: 19879807]

[52] Geraci JR, St Aubin DJ, Barker IK, *et al.* Mass mortality of harbor seals: pneumonia associated with influenza A virus. Science 1982; 215(4536): 1129-31.
[http://dx.doi.org/10.1126/science.7063847] [PMID: 7063847]

[53] Guo YJ, Jin FG, Wang P, Wang M, Zhu JM. Isolation of influenza C virus from pigs and experimental infection of pigs with influenza C virus. J Gen Virol 1983; 64(Pt 1): 177-82.
[http://dx.doi.org/10.1099/0022-1317-64-1-177] [PMID: 6296296]

[54] Brown IH, Harris PA, Alexander DJ. Serological studies of influenza viruses in pigs in Great Britain 1991-2. Epidemiol Infect 1995; 114(3): 511-20.
[http://dx.doi.org/10.1017/S0950268800052225] [PMID: 7781739]

[55] Yamaoka M, Hotta H, Itoh M, Homma M. Prevalence of antibody to influenza C virus among pigs in Hyogo Prefecture, Japan. J Gen Virol 1991; 72(Pt 3): 711-4.
[http://dx.doi.org/10.1099/0022-1317-72-3-711] [PMID: 1848603]

[56] Bodewes R, Morick D, de Mutsert G, *et al.* Recurring influenza B virus infections in seals. Emerg Infect Dis 2013; 19(3): 511-2.
[http://dx.doi.org/10.3201/eid1903.120965] [PMID: 23750359]

[57] Osterhaus AD, Rimmelzwaan GF, Martina BE, Bestebroer TM, Fouchier RA. Influenza B virus in seals. Science 2000; 288(5468): 1051-3.
[http://dx.doi.org/10.1126/science.288.5468.1051] [PMID: 10807575]

[58] Mehle A. Unusual influenza A viruses in bats. Viruses 2014; 6(9): 3438-49.
[http://dx.doi.org/10.3390/v6093438] [PMID: 25256392]

[59] Poole DS, Yú S, Caì Y, *et al.* Influenza A virus polymerase is a site for adaptive changes during experimental evolution in bat cells. J Virol 2014; 88(21): 12572-85.
[http://dx.doi.org/10.1128/JVI.01857-14] [PMID: 25142579]

[60] Chiapponi C, Faccini S, De Mattia A, *et al.* Detection of influenza D virus among swine and cattle, Italy. Emerg Infect Dis 2016; 22(2): 352-4.
[http://dx.doi.org/10.3201/eid2202.151439] [PMID: 26812282]

[61] Hause BM, Collin EA, Liu R, *et al.* Characterization of a novel influenza virus in cattle and Swine: proposal for a new genus in the Orthomyxoviridae family. MBio 2014; 5(2): e00031-14.
[http://dx.doi.org/10.1128/mBio.00031-14] [PMID: 24595369]

[62] Liang L, Deng G, Shi J, *et al.* Genetics, receptor binding, replication, and mammalian transmission of H4 avian influenza viruses isolated from live poultry markets in China. J Virol 2015; 90(3): 1455-69.
[http://dx.doi.org/10.1128/JVI.02692-15] [PMID: 26581996]

[63] Murakami S, Odagiri T, Melaku SK, *et al.* Influenza D virus infection in dromedary camels, Ethiopia. Emerg Infect Dis 2019; 25(6): 1224-6.
[http://dx.doi.org/10.3201/eid2506.181158] [PMID: 31107233]

[64] Suzuki Y, Nei M. Origin and evolution of influenza virus hemagglutinin genes. Mol Biol Evol 2002;

19(4): 501-9.
[http://dx.doi.org/10.1093/oxfordjournals.molbev.a004105] [PMID: 11919291]

[65] Scholtissek C. Source for influenza pandemics. Eur J Epidemiol 1994; 10(4): 455-8.
[http://dx.doi.org/10.1007/BF01719674] [PMID: 7843354]

[66] Widjaja L, Krauss SL, Webby RJ, Xie T, Webster RG. Matrix gene of influenza a viruses isolated from wild aquatic birds: ecology and emergence of influenza a viruses. J Virol 2004; 78(16): 8771-9.
[http://dx.doi.org/10.1128/JVI.78.16.8771-8779.2004] [PMID: 15280485]

[67] Ghedin E, Sengamalay NA, Shumway M, *et al.* Large-scale sequencing of human influenza reveals the dynamic nature of viral genome evolution. Nature 2005; 437(7062): 1162-6.
[http://dx.doi.org/10.1038/nature04239] [PMID: 16208317]

[68] Parvin JD, Moscona A, Pan WT, Leider JM, Palese P. Measurement of the mutation rates of animal viruses: influenza A virus and poliovirus type 1. J Virol 1986; 59(2): 377-83.
[http://dx.doi.org/10.1128/JVI.59.2.377-383.1986] [PMID: 3016304]

[69] Smith DB, Inglis SC. The mutation rate and variability of eukaryotic viruses: an analytical review. J Gen Virol 1987; 68(Pt 11): 2729-40.
[http://dx.doi.org/10.1099/0022-1317-68-11-2729] [PMID: 3316486]

[70] Brand C, Palese P. Sequential passage of influenza virus in embryonated eggs or tissue culture: emergence of mutants. Virology 1980; 107(2): 424-33.
[http://dx.doi.org/10.1016/0042-6822(80)90309-8] [PMID: 6256942]

[71] Holland J, Spindler K, Horodyski F, Grabau E, Nichol S, VandePol S. Rapid evolution of RNA genomes. Science 1982; 215(4540): 1577-85.
[http://dx.doi.org/10.1126/science.7041255] [PMID: 7041255]

[72] Gong LI, Suchard MA, Bloom JD. Stability-mediated epistasis constrains the evolution of an influenza protein. eLife 2013; 2: e00631.
[http://dx.doi.org/10.7554/eLife.00631] [PMID: 23682315]

[73] Kilbourne ED. Future influenza vaccines and the use of genetic recombinants. Bull World Health Organ 1969; 41(3): 643-5.
[PMID: 5309489]

[74] Moya A, Holmes EC, González-Candelas F. The population genetics and evolutionary epidemiology of RNA viruses. Nat Rev Microbiol 2004; 2(4): 279-88.
[http://dx.doi.org/10.1038/nrmicro863] [PMID: 15031727]

[75] Marshall N, Priyamvada L, Ende Z, Steel J, Lowen AC. Influenza virus reassortment occurs with high frequency in the absence of segment mismatch. PLoS Pathog 2013; 9(6): e1003421.
[http://dx.doi.org/10.1371/journal.ppat.1003421] [PMID: 23785286]

[76] Gao Q, Palese P. Rewiring the RNAs of influenza virus to prevent reassortment. Proc Natl Acad Sci USA 2009; 106(37): 15891-6.
[http://dx.doi.org/10.1073/pnas.0908897106] [PMID: 19805230]

[77] Li C, Hatta M, Watanabe S, Neumann G, Kawaoka Y. Compatibility among polymerase subunit proteins is a restricting factor in reassortment between equine H7N7 and human H3N2 influenza viruses. J Virol 2008; 82(23): 11880-8.
[http://dx.doi.org/10.1128/JVI.01445-08] [PMID: 18815312]

[78] Subbarao K, Webster RG, Kawaoka Y, Murphy BR. Are there alternative avian influenza viruses for generation of stable attenuated avian-human influenza A reassortant viruses? Virus Res 1995; 39(2-3): 105-18.
[http://dx.doi.org/10.1016/0168-1702(95)00082-8] [PMID: 8837878]

[79] Chen R, Holmes EC. The evolutionary dynamics of human influenza B virus. J Mol Evol 2008; 66(6): 655-63.
[http://dx.doi.org/10.1007/s00239-008-9119-z] [PMID: 18504518]

[80] Dudas G, Bedford T, Lycett S, Rambaut A. Reassortment between influenza B lineages and the emergence of a coadapted PB1-PB2-HA gene complex. Mol Biol Evol 2015; 32(1): 162-72. [http://dx.doi.org/10.1093/molbev/msu287] [PMID: 25323575]

[81] Belshaw R, Gardner A, Rambaut A, Pybus OG. Pacing a small cage: mutation and RNA viruses. Trends Ecol Evol 2008; 23(4): 188-93. [http://dx.doi.org/10.1016/j.tree.2007.11.010] [PMID: 18295930]

[82] Hutchinson EC, von Kirchbach JC, Gog JR, Digard P. Genome packaging in influenza A virus. J Gen Virol 2010; 91(Pt 2): 313-28. [http://dx.doi.org/10.1099/vir.0.017608-0] [PMID: 19955561]

[83] Garten RJ, Davis CT, Russell CA, *et al.* Antigenic and genetic characteristics of swine-origin 2009 A(H1N1) influenza viruses circulating in humans. Science 2009; 325(5937): 197-201. [http://dx.doi.org/10.1126/science.1176225] [PMID: 19465683]

[84] Webster RG, Bean WJ, Gorman OT, Chambers TM, Kawaoka Y. Evolution and ecology of influenza A viruses. Microbiol Rev 1992; 56(1): 152-79. [http://dx.doi.org/10.1128/MR.56.1.152-179.1992] [PMID: 1579108]

[85] Shortridge KF. Pandemic influenza: a zoonosis? Semin Respir Infect 1992; 7(1): 11-25. [PMID: 1609163]

[86] Bouvier NM, Palese P. The biology of influenza viruses. Vaccine 2008; 4(Suppl 4): D49-53. [http://dx.doi.org/10.1016/j.vaccine.2008.07.039] [PMID: 19230160]

[87] Ren XW, Ju LW, Yang JX, *et al.* Antigenic and genetic variation in the hemagglutinins of H1N1 and H3N2 human influenza a viruses in the Shanghai area from 2005 to 2008. J Med Virol 2011; 83(7): 1113-20. [http://dx.doi.org/10.1002/jmv.22078] [PMID: 21520138]

[88] Cherry JL, Lipman DJ, Nikolskaya A, Wolf YI. Evolutionary dynamics of N-glycosylation sites of influenza virus hemagglutinin. PLoS Curr 2009; 1: RRN1001. [http://dx.doi.org/10.1371/currents.RRN1001] [PMID: 20025194]

[89] Hay AJ, Gregory V, Douglas AR, Lin YP. The evolution of human influenza viruses. Philos Trans R Soc Lond B Biol Sci 2001; 356(1416): 1861-70. [http://dx.doi.org/10.1098/rstb.2001.0999] [PMID: 11779385]

[90] Tate MD, Job ER, Deng YM, Gunalan V, Maurer-Stroh S, Reading PC. Playing hide and seek: how glycosylation of the influenza virus hemagglutinin can modulate the immune response to infection. Viruses 2014; 6(3): 1294-316. [http://dx.doi.org/10.3390/v6031294] [PMID: 24638204]

[91] Wei CJ, Boyington JC, Dai K, *et al.* Cross-neutralization of 1918 and 2009 influenza viruses: role of glycans in viral evolution and vaccine design. Sci Transl Med 2010; 2(24): 24ra21. [http://dx.doi.org/10.1126/scitranslmed.3000799] [PMID: 20375007]

[92] Sun S, Wang Q, Zhao F, Chen W, Li Z. Glycosylation site alteration in the evolution of influenza A (H1N1) viruses. PLoS One 2011; 6(7): e22844. [http://dx.doi.org/10.1371/journal.pone.0022844] [PMID: 21829533]

[93] Gambaryan AS, Robertson JS, Matrosovich MN. Effects of egg-adaptation on the receptor-binding properties of human influenza A and B viruses. Virology 1999; 258(2): 232-9. [http://dx.doi.org/10.1006/viro.1999.9732] [PMID: 10366560]

[94] Reid AH, Taubenberger JK, Fanning TG. Evidence of an absence: the genetic origins of the 1918 pandemic influenza virus. Nat Rev Microbiol 2004; 2(11): 909-14. [http://dx.doi.org/10.1038/nrmicro1027] [PMID: 15494747]

[95] Taubenberger JK, Reid AH, Krafft AE, Bijwaard KE, Fanning TG. Initial genetic characterization of the 1918 "Spanish" influenza virus. Science 1997; 275(5307): 1793-6.

[http://dx.doi.org/10.1126/science.275.5307.1793] [PMID: 9065404]

[96] Kawaoka Y, Krauss S, Webster RG. Avian-to-human transmission of the PB1 gene of influenza A viruses in the 1957 and 1968 pandemics. J Virol 1989; 63(11): 4603-8. [http://dx.doi.org/10.1128/JVI.63.11.4603-4608.1989] [PMID: 2795713]

[97] Lindstrom SE, Cox NJ, Klimov A. Genetic analysis of human H2N2 and early H3N2 influenza viruses, 1957-1972: evidence for genetic divergence and multiple reassortment events. Virology 2004; 328(1): 101-19. [http://dx.doi.org/10.1016/j.virol.2004.06.009] [PMID: 15380362]

[98] Frost WH. Statistics of influenza morbidity: With special reference to certain factors in case incidence and case fatality. Public Health Reports (1896-1970) 1920; 35(11): 584-97. [Available from: https://www.jstor.org/stable/4575511]

[99] Kobasa D, Jones SM, Shinya K, *et al.* Aberrant innate immune response in lethal infection of macaques with the 1918 influenza virus. Nature 2007; 445(7125): 319-23. [http://dx.doi.org/10.1038/nature05495] [PMID: 17230189]

[100] Neumann G, Watanabe T, Ito H, *et al.* Generation of influenza A viruses entirely from cloned cDNAs. Proc Natl Acad Sci USA 1999; 96(16): 9345-50. [http://dx.doi.org/10.1073/pnas.96.16.9345] [PMID: 10430945]

[101] Nelson MI, Worobey M. Origins of the 1918 pandemic: Revisiting the swine "mixing vessel" hypothesis. Am J Epidemiol 2018; 187(12): 2498-502. [http://dx.doi.org/10.1093/aje/kwy150] [PMID: 30508193]

[102] Scholtissek C, Rohde W, Von Hoyningen V, Rott R. On the origin of the human influenza virus subtypes H2N2 and H3N2. Virology 1978; 87(1): 13-20. [http://dx.doi.org/10.1016/0042-6822(78)90153-8] [PMID: 664248]

[103] Nakajima K, Desselberger U, Palese P. Recent human influenza A (H1N1) viruses are closely related genetically to strains isolated in 1950. Nature 1978; 274(5669): 334-9. [http://dx.doi.org/10.1038/274334a0] [PMID: 672956]

[104] Dawood FS, Jain S, Finelli L, *et al.* Emergence of a novel swine-origin influenza A (H1N1) virus in humans. N Engl J Med 2009; 360(25): 2605-15. [http://dx.doi.org/10.1056/NEJMoa0903810] [PMID: 19423869]

[105] Morens DM, Taubenberger JK, Fauci AS. The persistent legacy of the 1918 influenza virus. N Engl J Med 2009; 361(3): 225-9. [http://dx.doi.org/10.1056/NEJMp0904819] [PMID: 19564629]

[106] Mubareka S, Palese P. Influenza virus: The biology of a changing virus. In: Rappuoli R, Del Giudice G, Eds. Influenza Vaccines for the Future. 2nd ed. Basel: Springer 2008; pp. 3-26. [http://dx.doi.org/10.1007/978-3-7643-8371-8_2]

[107] WHO. Emergencies preparedness, response. Pandemic (H1N1) 2009 - update 92. 2010. [Available from: http://www.who.int/csr/don/2010_03_19/en/index.html]

[108] Dawood FS, Iuliano AD, Reed C, *et al.* Estimated global mortality associated with the first 12 months of 2009 pandemic influenza A H1N1 virus circulation: a modelling study. Lancet Infect Dis 2012; 12(9): 687-95. [http://dx.doi.org/10.1016/S1473-3099(12)70121-4] [PMID: 22738893]

[109] Simonsen L, Spreeuwenberg P, Lustig R, *et al.* GLaMOR Collaborating Teams. Global mortality estimates for the 2009 Influenza Pandemic from the GLaMOR project: a modeling study. PLoS Med 2013; 10(11): e1001558. [http://dx.doi.org/10.1371/journal.pmed.1001558] [PMID: 24302890]

[110] CDC. First global estimates of 2009 H1N1 pandemic mortality released by CDC-Led collaboration. 2012. [Available from: http://www.cdc.gov/flu/spotlights/pandemic-global-estimates.htm]

[111] Bautista E, Chotpitayasunondh T, Gao Z, *et al.* Clinical aspects of pandemic 2009 influenza A (H1N1) virus infection. N Engl J Med 2010; 362(18): 1708-19. [http://dx.doi.org/10.1056/NEJMra1000449] [PMID: 20445182]

[112] WHO. Emergencies preparedness, response. 2009. [Available from: http://www.who.int/csr/disease/swineflu/notes /h1n1_clinical_features_20091016/en/]

[113] Krammer F, Palese P. Influenza virus hemagglutinin stalk-based antibodies and vaccines. Curr Opin Virol 2013; 3(5): 521-30. [http://dx.doi.org/10.1016/j.coviro.2013.07.007] [PMID: 23978327]

[114] Nachbagauer R, Salaun B, Stadlbauer D, *et al.* Pandemic influenza virus vaccines boost hemagglutinin stalk-specific antibody responses in primed adult and pediatric cohorts. NPJ Vacc 2019; 4: 51. [http://dx.doi.org/10.1038/s41541-019-0147-z] [PMID: 31839997]

[115] Palese P, Wang TT. Why do influenza virus subtypes die out? A hypothesis. MBio 2011; 2(5): e00150-11. [http://dx.doi.org/10.1128/mBio.00150-11] [PMID: 21878571]

[116] Kilbourne ED, Cerini CP, Khan MW, Mitchell JW Jr, Ogra PL. Immunologic response to the influenza virus neuraminidase is influenced by prior experience with the associated viral hemagglutinin. I. Studies in human vaccinees. J Immunol 1987; 138(9): 3010-3. [PMID: 3571981]

[117] Schweiger B, Zadow I, Heckler R. Antigenic drift and variability of influenza viruses. Med Microbiol Immunol (Berl) 2002; 191(3-4): 133-8. [http://dx.doi.org/10.1007/s00430-002-0132-3] [PMID: 12458347]

[118] Paul Glezen W, Schmier JK, Kuehn CM, Ryan KJ, Oxford J. The burden of influenza B: a structured literature review. Am J Public Health 2013; 103(3): e43-51. [http://dx.doi.org/10.2105/AJPH.2012.301137] [PMID: 23327249]

[119] Bedford T, Riley S, Barr IG, *et al.* Global circulation patterns of seasonal influenza viruses vary with antigenic drift. Nature 2015; 523(7559): 217-20. [http://dx.doi.org/10.1038/nature14460] [PMID: 26053121]

[120] Rambaut A, Pybus OG, Nelson MI, Viboud C, Taubenberger JK, Holmes EC. The genomic and epidemiological dynamics of human influenza A virus. Nature 2008; 453(7195): 615-9. [http://dx.doi.org/10.1038/nature06945] [PMID: 18418375]

[121] Russell CA, Jones TC, Barr IG, *et al.* The global circulation of seasonal influenza A (H3N2) viruses. Science 2008; 320(5874): 340-6. [http://dx.doi.org/10.1126/science.1154137] [PMID: 18420927]

[122] Lemey P, Rambaut A, Bedford T, *et al.* Unifying viral genetics and human transportation data to predict the global transmission dynamics of human influenza H3N2. PLoS Pathog 2014; 10(2): e1003932. [http://dx.doi.org/10.1371/journal.ppat.1003932] [PMID: 24586153]

[123] Nerome R, Hiromoto Y, Sugita S, *et al.* Evolutionary characteristics of influenza B virus since its first isolation in 1940: dynamic circulation of deletion and insertion mechanism. Arch Virol 1998; 143(8): 1569-83. [http://dx.doi.org/10.1007/s007050050399] [PMID: 9739335]

[124] Capua I, Alexander DJ. Avian influenza and human health. Acta Trop 2002; 83(1): 1-6. [http://dx.doi.org/10.1016/S0001-706X(02)00050-5] [PMID: 12062786]

[125] CDC. Examples of human infections with avian influenza A viruses with possible limited, non-sustained human-to-human transmission. 2020. [Available from: https://www.cdc.gov/flu/avianflu/h5n1-human-infections.htm]

[126] Claas EC, Osterhaus AD, van Beek R, *et al.* Human influenza A H5N1 virus related to a highly

pathogenic avian influenza virus. Lancet 1998; 351(9101): 472-7. [http://dx.doi.org/10.1016/S0140-6736(97)11212-0] [PMID: 9482438]

[127] Subbarao K, Klimov A, Katz J, *et al.* Characterization of an avian influenza A (H5N1) virus isolated from a child with a fatal respiratory illness. Science 1998; 279(5349): 393-6. [http://dx.doi.org/10.1126/science.279.5349.393] [PMID: 9430591]

[128] Guan Y, Peiris JS, Lipatov AS, *et al.* Emergence of multiple genotypes of H5N1 avian influenza viruses in Hong Kong SAR. Proc Natl Acad Sci USA 2002; 99(13): 8950-5. [http://dx.doi.org/10.1073/pnas.132268999] [PMID: 12077307]

[129] Jiao P, Song H, Liu X, *et al.* Pathogenicity, transmission and antigenic variation of H5N1 highly pathogenic avian influenza viruses. Front Microbiol 2016; 7: 635. [http://dx.doi.org/10.3389/fmicb.2016.00635] [PMID: 27199961]

[130] Smith GJ, Fan XH, Wang J, *et al.* Emergence and predominance of an H5N1 influenza variant in China. Proc Natl Acad Sci USA 2006; 103(45): 16936-41. [http://dx.doi.org/10.1073/pnas.0608157103] [PMID: 17075062]

[131] Ducatez MF, Olinger CM, Owoade AA, *et al.* Avian flu: multiple introductions of H5N1 in Nigeria. Nature 2006; 442(7098): 37. [http://dx.doi.org/10.1038/442037a] [PMID: 16823443]

[132] WHO. Cumulative number of confirmed human cases for avian influenza A(H5N1) reported to World Health Organization, 2003-2020. [Available from: https://www.who.int/influenza/human_animal_interface/2020_01_20_tableH5N1.pdf?ua=1]

[133] CDC. Highly pathogenic asian avian influenza A(H5N1) in people. 2020. [Available from: http://www.cdc.gov/flu/avianflu/h5n1-people.htm]

[134] Bi Y, Mei K, Shi W, *et al.* Two novel reassortants of avian influenza A (H5N6) virus in China. J Gen Virol 2015; 96(Pt 5): 975-81. [http://dx.doi.org/10.1099/vir.0.000056] [PMID: 25604926]

[135] WHO. Influenza. Monthly risk assessment summary. 2020. [Available from: http://www.who.int/influenza/human_animal_interface/HAI_Risk_Assessment/en]

[136] Tate MD. Highly pathogenic avian H5N8 influenza viruses: should we be concerned? Virulence 2018; 9(1): 20-1. [http://dx.doi.org/10.1080/21505594.2017.1386832] [PMID: 28968185]

[137] Influenza FAO. FAO. Influenza: Food and Agriculture Organization of the United Nations. H7N9 Situation Update 2019. [Available from: www.fao.org/ag/againfo/programmes/en/empres/H7N9/situation_update.html]

[138] Wang C, Luo J, Wang J, *et al.* Novel human H7N9 influenza virus in China. Integr Zool 2014; 9(3): 372-5. [http://dx.doi.org/10.1111/1749-4877.12047] [PMID: 24952971]

[139] Gao R, Cao B, Hu Y, *et al.* Human infection with a novel avian-origin influenza A (H7N9) virus. N Engl J Med 2013; 368(20): 1888-97. [http://dx.doi.org/10.1056/NEJMoa1304459] [PMID: 23577628]

[140] Zhang Y, Zhang Q, Kong H, *et al.* H5N1 hybrid viruses bearing 2009/H1N1 virus genes transmit in guinea pigs by respiratory droplet. Science 2013; 340(6139): 1459-63. [http://dx.doi.org/10.1126/science.1229455] [PMID: 23641061]

[141] Kucharski AJ, Mills HL, Donnelly CA, Riley S. Transmission potential of influenza A(H7N9) virus, China, 2013-2014. Emerg Infect Dis 2015; 21(5): 852-5. [http://dx.doi.org/10.3201/eid2105.141137] [PMID: 25897624]

[142] WHO. Recommended composition of influenza virus vaccines for use in the 2017 southern hemisphere influenza season. 2016. [Available from: https://www.who.int/influenza/vaccines/virus

/recommendations/201609_recommendation.pdf?ua=1]

[143] Sun Y, Liu J. H9N2 influenza virus in China: a cause of concern. Protein Cell 2015; 6(1): 18-25. [http://dx.doi.org/10.1007/s13238-014-0111-7] [PMID: 25384439]

[144] Khan SU, Anderson BD, Heil GL, Liang S, Gray GC. A Systematic review and meta-analysis of the seroprevalence of influenza A(H9N2) infection among humans. J Infect Dis 2015; 212(4): 562-9. [http://dx.doi.org/10.1093/infdis/jiv109] [PMID: 25712969]

[145] Lopez-Martinez I, Balish A, Barrera-Badillo G, *et al.* Highly pathogenic avian influenza A(H7N3) virus in poultry workers, Mexico, 2012. Emerg Infect Dis 2013; 19(9): 1531-4. [http://dx.doi.org/10.3201/eid1909.130087] [PMID: 23965808]

[146] Chen H, Yuan H, Gao R, *et al.* Clinical and epidemiological characteristics of a fatal case of avian influenza A H10N8 virus infection: a descriptive study. Lancet 2014; 383(9918): 714-21. [http://dx.doi.org/10.1016/S0140-6736(14)60111-2] [PMID: 24507376]

[147] Wang G, Deng G, Shi J, *et al.* H6 influenza viruses pose a potential threat to human health. J Virol 2014; 88(8): 3953-64. [http://dx.doi.org/10.1128/JVI.03292-13] [PMID: 24501418]

[148] Yuan J, Zhang L, Kan X, *et al.* Origin and molecular characteristics of a novel 2013 avian influenza A(H6N1) virus causing human infection in Taiwan. Clin Infect Dis 2013; 57(9): 1367-8. [http://dx.doi.org/10.1093/cid/cit479] [PMID: 23881153]

[149] Zhang T, Bi Y, Tian H, *et al.* Human infection with influenza virus A(H10N8) from live poultry markets, China, 2014. Emerg Infect Dis 2014; 20(12): 2076-9. [http://dx.doi.org/10.3201/eid2012.140911] [PMID: 25425075]

[150] Yoo SJ, Kwon T, Lyoo YS. Challenges of influenza A viruses in humans and animals and current animal vaccines as an effective control measure. Clin Exp Vaccine Res 2018; 7(1): 1-15. [http://dx.doi.org/10.7774/cevr.2018.7.1.1] [PMID: 29399575]

[151] Puzelli S, Rossini G, Facchini M, *et al.* Human infection with highly pathogenic A(H7N7) avian influenza virus, Italy, 2013. Emerg Infect Dis 2014; 20(10): 1745-9. [http://dx.doi.org/10.3201/eid2010.140512] [PMID: 25271444]

[152] WHO. Emergencies preparedness, response. Influenza: Human infection with avian influenza A(H7N4) virus – China. 2018. [Available from: https://www.who.int/csr/don/22-february-20-8-ah7n4-china/en/]

[153] Kendal AP, Goldfield M, Noble GR, Dowdle WR. Identification and preliminary antigenic analysis of swine influenza-like viruses isolated during an influenza outbreak at Fort Dix, New Jersey. J Infect Dis 1977; 136 (Suppl.): S381-5. [http://dx.doi.org/10.1093/infdis/136.Supplement_3.S381] [PMID: 606762]

[154] Lewis NS, Anderson TK, Kitikoon P, Skepner E, Burke DF, Vincent AL. Substitutions near the hemagglutinin receptor-binding site determine the antigenic evolution of influenza A H3N2 viruses in U.S. swine. J Virol 2014; 88(9): 4752-63. [http://dx.doi.org/10.1128/JVI.03805-13] [PMID: 24522915]

[155] CDC. Influenza A (H3N2) variant virus. 2020. [Available from: http://www.cdc.gov/flu/swineflu/h3n2v-cases.htm]

[156] Oxford JS, Öberg B. Conquest of viral disease: A topical review of drug and vaccines Amsterdam. Elsevier 1985.

[157] Cox NJ, Kitame F, Kendal AP, Maassab HF, Naeve C. Identification of sequence changes in the cold-adapted, live attenuated influenza vaccine strain, A/Ann Arbor/6/60 (H2N2). Virology 1988; 167(2): 554-67. [http://dx.doi.org/10.1016/0042-6822(88)90118-3] [PMID: 2974219]

[158] Herlocher ML, Maassab HF, Webster RG. Molecular and biological changes in the cold-adapted

"master strain" A/AA/6/60 (H2N2) influenza virus. Proc Natl Acad Sci USA 1993; 90(13): 6032-6. [http://dx.doi.org/10.1073/pnas.90.13.6032] [PMID: 8327480]

[159] Reeve P, Almond JW, Felsenreich V, Pibermann M, Maassab HF. Studies with a cold-recombinant A/Victoria/3/75 (H3N2) virus. I. biologic, genetic, and biochemical characterization. J Infect Dis 1980; 142(6): 850-6. [http://dx.doi.org/10.1093/infdis/142.6.850] [PMID: 7462696]

[160] Larionova N, Kiseleva I, Isakova I, Litvinova O, Klimov A, Rudenko L. Naturally occurring temperature-sensitive strains of influenza B virus. In: Cox N, Couch RB, Belshe RB, Eds. Proceedings of IVW-2004 Conference. Lisbon, Portugal 2004; pp. 92-7.

[161] Chu CM, Tian SF, Ren GF, Zhang YM, Zhang LX, Liu GQ. Occurrence of temperature-sensitive influenza A viruses in nature. J Virol 1982; 41(2): 353-9. [http://dx.doi.org/10.1128/JVI.41.2.353-359.1982] [PMID: 7077746]

[162] Yamaya M, Nishimura H, Lusamba Kalonji N, *et al.* Effects of high temperature on pandemic and seasonal human influenza viral replication and infection-induced damage in primary human tracheal epithelial cell cultures. Heliyon 2019; 5(2): e01149. [http://dx.doi.org/10.1016/j.heliyon.2019.e01149] [PMID: 30839917]

[163] Kiseleva IV, Klimov AI, Grigor'eva EP, Larionova NV, Aleksandrova GI, Rudenko LG. Genetic and phenotypic analysis of heterogeneous population of a cold-adapted donor of the A/Leningrad/134/17/57 (H2N2) attenuation and of the donor-based reassortant influenza vaccine strains. Vopr Virusol 2005; 50(2): 14-8. [PMID: 15881391]

[164] Maassab HF, DeBorde DC. Development and characterization of cold-adapted viruses for use as live virus vaccines. Vaccine 1985; 3(5): 355-69. [http://dx.doi.org/10.1016/0264-410X(85)90124-0] [PMID: 3909681]

[165] Oxford JS, Corcoran T, Schild GC. Naturally occurring temperature-sensitive influenza A viruses of the H1N1 and H3N2 subtypes. J Gen Virol 1980; 48(Pt 2): 383-9. [http://dx.doi.org/10.1099/0022-1317-48-2-383] [PMID: 7400776]

[166] Polezhaev FI, Aleksandrova GI. Isolation of temperature-sensitive strains of the influenza virus in the epidemic caused by the A/Victoria virus in 1975-1976. Vopr Virusol 1979; (4): 430. [PMID: 483779]

[167] Herlocher ML, Clavo AC, Maassab HF. Sequence comparisons of A/AA/6/60 influenza viruses: mutations which may contribute to attenuation. Virus Res 1996; 42(1-2): 11-25. [http://dx.doi.org/10.1016/0168-1702(96)01292-0] [PMID: 8806171]

[168] Kiseleva IV, Larionova NV, Voeten JT, *et al.* Leading role of genes coding polymerase complex in attenuation of domestic donor viruses for A and B live influenza vaccine. Zh Mikrobiol Epidemiol Immunobiol 2010; (6): 41-7. [PMID: 21381376]

[169] Kiseleva IV, Voeten JT, Teley LC, *et al.* PB2 and PA genes control the expression of the temperature-sensitive phenotype of cold-adapted B/USSR/60/69 influenza master donor virus. J Gen Virol 2010; 91(Pt 4): 931-7. [http://dx.doi.org/10.1099/vir.0.017996-0] [PMID: 20007357]

[170] Massin P, van der Werf S, Naffakh N. Residue 627 of PB2 is a determinant of cold sensitivity in RNA replication of avian influenza viruses. J Virol 2001; 75(11): 5398-404. [http://dx.doi.org/10.1128/JVI.75.11.5398-5404.2001] [PMID: 11333924]

[171] Giles BM, Ross TM. A computationally optimized broadly reactive antigen (COBRA) based H5N1 VLP vaccine elicits broadly reactive antibodies in mice and ferrets. Vaccine 2011; 29(16): 3043-54. [http://dx.doi.org/10.1016/j.vaccine.2011.01.100] [PMID: 21320540]

[172] McCauley JW, Penn CR. The critical cut-off temperature of avian influenza viruses. Virus Res 1990;

17(3): 191-8.
[http://dx.doi.org/10.1016/0168-1702(90)90064-I] [PMID: 2077784]

[173] Larionova NV, Kiseleva IV, Rudenko LG. Evolution of influenza viruses based on sensitivity to temperature of replication. J Microb Epidemiol Immunobiol 6(3): 47-55.

[174] Larionova NV, Kiseleva IV, Isakova IN, Litvinova OM, Rudenko LG. Phenotyte of epidemic influenza B virus strains isolated in different years. Vopr Virusol 2006; 51(5): 38-41. [PMID: 17087064]

CHAPTER 4

Influenza Prophylaxis and Treatment

Abstract: Each year, influenza causes a significant acute respiratory disease burden. Also, influenza pandemics periodically occur. Annual vaccination is the best tool for influenza prevention, but its effectiveness can vary from year to year. Antiviral treatment shortens the clinical course if administered within the first hours of the onset of the disease. Neuraminidase inhibitors (oseltamivir, zanamivir, and peramivir) and polymerase inhibitors (baloxavir) are currently used as antivirals. Limited specificity of conventional vaccines and drug resistance of circulating viruses reduce the effectiveness of prophylaxis and treatment, requiring the development of new broad-spectrum drugs.

Keywords: Cross-protectivity, Direct-acting antivirals, Drug resistance, Inactivated influenza vaccines, Indirectly acting antivirals, Live attenuated influenza vaccines, Next-generation vaccines, Synthetic vaccines.

INTRODUCTION

In the first quarter of the 21st century, influenza epidemics and pandemics continue to be a threat to humankind. Of course, seasonal influenza remains one of the major public health threats. However, concerns regarding the possibility of pandemic spread of novel influenza strains have led to increased interest in new influenza vaccines and drug development.

Isolation of the first influenza virus led to the development of the first generation of inactivated and live influenza vaccines. A hundred years after the influenza virus was first isolated, effective influenza vaccines are still an important influenza prevention tool and are the key element in influenza control [1]. Vaccination can reduce illness and lessen the severity of infection, especially in high-risk groups [2]. Innovative strategies, such as recombinant technologies, are currently being studied to improve the immunological response to vaccination [3] and develop universal vaccines [4].

The current strategy for the fight against influenza infection involves the development of a global epidemiological surveillance network [5], methods of rapid diagnosis, preventive vaccination of the population and the use of antivirals

Irina Kiseleva and Natalie Larionova

[6]. The recommendations of the European Center for Disease Control emphasize that vaccination is the main, most effective method of preventing the severe and complicated course of influenza infection, even if the vaccine is less effective than expected [7].

Disease management through antivirals is also an important component of the current intervention strategy to prevent and treat this infection. Antiviral treatment shortens the clinical course when given within two days from the onset of the disease. Antiviral drugs have been approved for treatment or prophylaxis of influenza A and B, caused by circulating influenza virus strains [8]. They are necessary in cases of diseases of non-immunized persons, and when the composition of the vaccine strains does not correspond to the viruses that caused the epidemic/pandemic.

Approved antiviral drugs for influenza fall into three classes, M2 channel inhibitors, neuraminidase inhibitors, and polymerase inhibitors [9 - 11]. Unfortunately, drug resistance of circulating viruses reduces the effectiveness of influenza treatment with approved antivirals and dictates the need for the development of new influenza antivirals.

PAST, PRESENT, AND FUTURE OF INFLUENZA VACCINES

Influenza vaccines may be classified into two basic types, namely live attenuated (LAIV) and inactivated (IIV) vaccines. LAIV can replicate in the susceptible hosts and stimulate an immune response by mimicking natural disease. The inactivated vaccine candidate contains the whole virus or parts of it; it is killed by special treatment and cannot replicate. Unlike LAIV, IIV typically requires multiple doses to ensure a sufficient immune response.

Besides, several subtypes of influenza vaccines exist. All of them will be discussed below in other sections.

Substrates for Influenza Vaccine Development

In the 1930s, the first influenza viruses were isolated in ferrets [12] and mice [13 - 15]. In 1938, a Russian live attenuated vaccine was developed by Smorodintsev in mice which were the most employed animals in virology [14]. This vaccine was developed by passing of the influenza virus through mice based on a genius Louis Pasteur's method of attenuation by sequential passing of the pathogen in another host. This mouse-derived LAIV was administered intranasally and used for influenza prophylaxis in adults in the USSR until 1947 (Fig. **1**). The following Fig. (**1**) displays the most important milestones in the development of influenza vaccines.

During the same period, another mice-derived influenza vaccine was developed in the US. It was a vaccine obtained directly from mouse lungs, or the mouse-derived virus was introduced into the so-called “tissue culture medium” which consisted of a minced chicken embryo [16]. Like the intranasal Russian vaccine, this preparation also contained live virus; however, it was administered subcutaneously and intradermally, and therefore can be considered as the first prototype of the inactivated influenza vaccine (IIV).

When chicken embryos were discovered as an alternative system for isolation of different viruses [17], a new era of virus isolation, including the influenza virus, began. After 1947, mice as the substrate for Russian LAIV development were completely replaced with embryonated chicken eggs (herein simply referred to as “eggs”). The US IIV and LAIV technology processes were also employed for egg-based vaccine production.

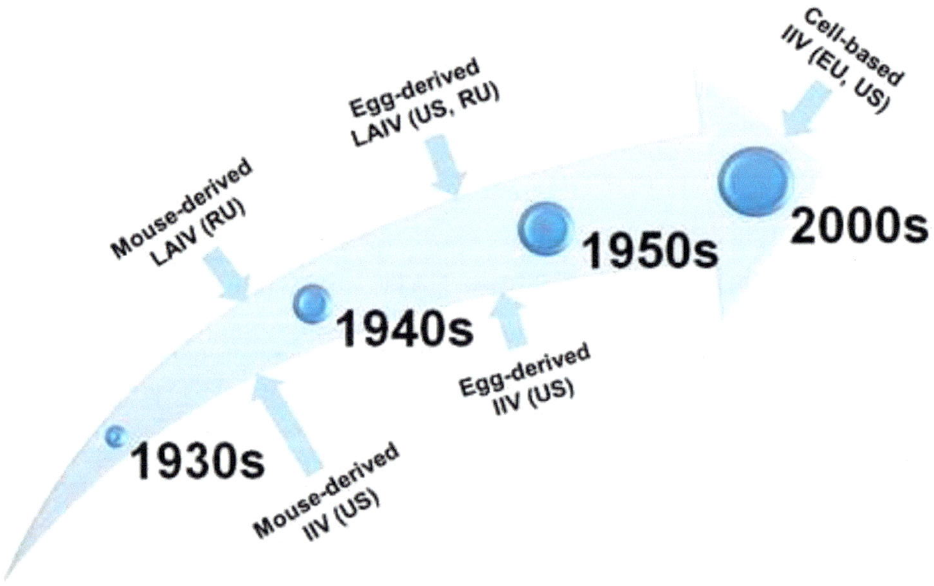

Fig. (1). Schematic evolution of influenza vaccine strategy based on the substrate for cultivation of influenza virus.

Today, three parallel approaches for influenza vaccine production are being used by vaccine manufacturers: (**i**) the production of conventional influenza vaccine in eggs, (**ii**) the production of the conventional influenza vaccine in cell culture, and the most recent technology is (**iii**) the production of so-called synthetic vaccines (Fig. **2**).

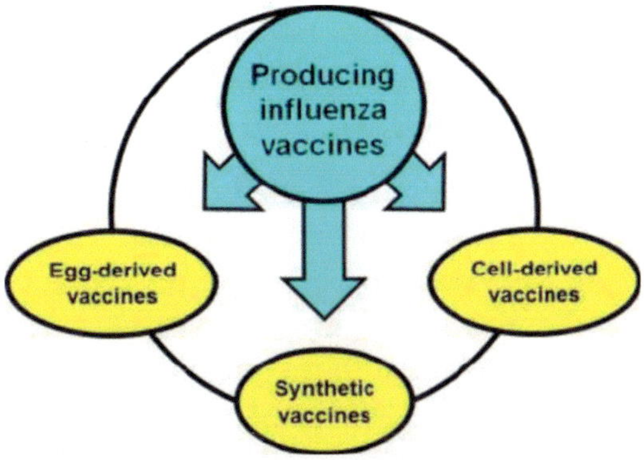

Fig. (2). Conventional and novel platforms for influenza vaccine production.

Conventional Egg-Derived Influenza Vaccines

The majority of influenza vaccines that are currently commercialized by licensed manufacturers are grown in eggs. "This is a labor-intensive process as each egg must be sterilized, candled, inoculated with the virus, and incubated before harvesting small volumes of allantoic fluid from each egg and pooling before purification" [18]. For each dose of inactivated vaccine, at least 1-2 eggs are required. For 10-15 doses of a live vaccine, the same number of eggs (1-2) are required. Therefore, millions of embryonated chicken eggs are required each year for seasonal vaccine production. In the event of a pandemic situation, their number will progressively increase.

Moreover, for a long time, eggs were the only WHO-approved substrate for influenza virus isolation, despite that in the 1990s, it was already shown that egg-derived variants of the influenza virus are often antigenically distinct from the original virus [19 - 22].

In addition, although cell culture has steadily replaced the use of eggs for the primary isolation of viruses from clinical specimens, candidate vaccine viruses must still be isolated directly in eggs according to current regulatory requirements [23]. Influenza vaccines for human use are licensed by national regulatory authorities, in which a product manufactured from influenza viruses is isolated and propagated exclusively in eggs; thus, cell culture isolates have been unacceptable for this purpose [24]. Manufacturers are not allowed to develop vaccine candidates based on cell-derived strains.

The limited availability of egg isolates, particularly of recent H3N2 viruses which generally grow poorly in eggs, may lead to serious problems in the vaccine virus selection process; timely recommendations for the use of H3N2 influenza vaccine candidates are required.

All the above problems raise a question regarding the further widespread use of eggs in vaccine manufacturing. As an alternative, significant efforts to increase the availability of egg isolates should be made [25].

Egg-derived live attenuated influenza vaccines. Humankind has benefited from live attenuated vaccines for more than two centuries. Louis Pasteur was the first scientist to make a crucial discovery by creating a vaccine in a laboratory setting [26]. He has found that by providing serial passage to a virus through a new species, the virus becomes adapted to that species and less adapted to its original host. In other words, the passaged virus becomes attenuated for its original host. As a result, the virulence of the passaged virus gets decreased for the original host. By passaging the virus through rabbits, Pasteur made the rabies virus less dangerous to human hosts, while still giving the body enough information to develop a proper immune response. Dozens of different vaccines, including LAIV, were created using the genius method of attenuation discovered by Louis Pasteur.

Mouse-derived influenza vaccine was replaced with the egg-derived vaccine. In 1957, during the Asian H2N2 pandemic, A/Singapore/1/57-like strain was isolated in Leningrad (the USSR) by employees of the Institute of Experimental Medicine (IEM). The isolate ID number 134 was designated as A/Leningrad/134/57 (H2N2) virus. The lab journal where this isolate was recorded is kept carefully in the Archive of IEM (Fig. **3**).

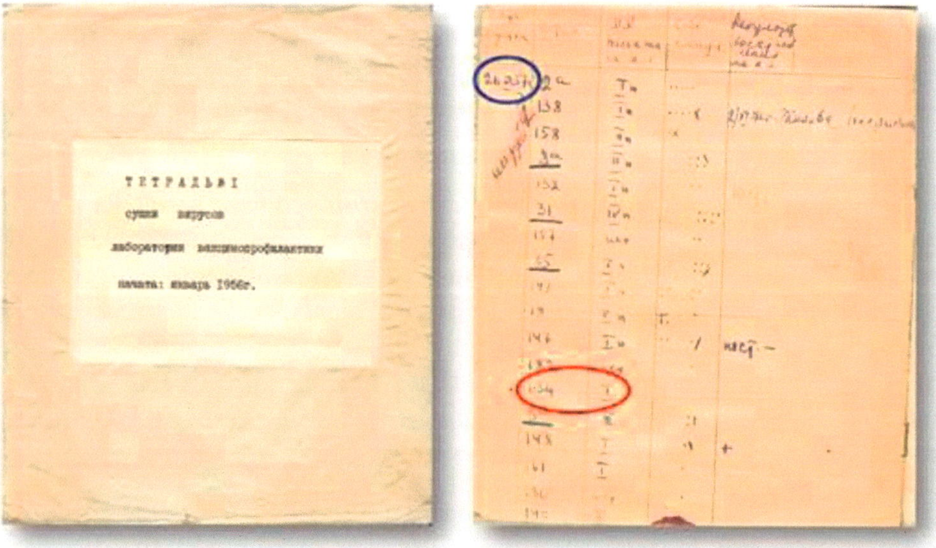

Fig. (3). Start the development of Russian LAIV in a lab. Left panel - Lab journal N°1 dating from January 1956, title page. Right panel - isolation of A/Leningrad/134/57 (A/Singapore/1/57-like) WT virus in 1957 at the Asian pandemic. Blue circle – date of isolation, red circle: 134 – number of isolate, I – number of passages at isolation. Photographs from the Archive of IEM, St Petersburg, Russia.

A/Leningrad/134/57 (H2N2) WT virus was passaged in eggs 21 times at the optimal temperature of 33ºC. The resulting virus became attenuated for humans [27] and was used in the USSR as pandemic H2N2 LAIV for healthy adults. However, children were not immunized with this vaccine because of its reactogenicity that was progressively increased with decreasing age of the children to be vaccinated [27].

In 1965, for additional attenuation, A/Leningrad (H2N2) vaccine virus was carried through 17 passages in eggs at the low temperature of 25ºC by Russian scientists Galina I. Alexandrova and Anatoly A. Smorodintsev [27]. The resulting virus obtained dramatically diminished multiplication capacity at elevated temperature of 38°C-40°C (*ts*, temperature-sensitive phenotype) and intensive replication capacity at a low temperature of 25ºC (*ca,* cold-adapted phenotype) and was safe and well-tolerated for children under the age of 1-2 years old. At the end of the Asian pandemic (1961–1967), this cryophilic strain was designated as A/Leningrad/134/17/57 (H2N2) and was used as a live pediatric vaccine for children over one-year-old. Later, the A/Leningrad/134/17/57 (H2N2) strain began to be used as master donor virus (MDV) for Russian cold-adapted reassortant LAIV.

In 1967, another great virologist, Hunein F. Maassab, published a similar article on an American live attenuated vaccine based on cold-adapted A/Ann Arbor/6/60 (H2N2) Asian pandemic strain (A/Singapore/1/57-like) (Fig. **4**) [28].

Passage history of the first cold-adapted version of A/Ann Arbor/6/60 (H2N2) differs from that of A/Leningrad/134/17/57 (H2N2). A/Leningrad/134/57 (H2N2) was isolated from nasal swabs in eggs, and all step-wise passages at optimal (33°C) and low (25°C) temperatures were performed in eggs, with 39 passages in total [27]. In contrast, A/Ann Arbor/6/60 (H2N2) was initially isolated from throat washings in primary chick kidney (PCK) cells at 36°C. It was undertaken 45 times in PCK, adapted to mouse lung through 15 intranasal passages and then back again to 10 passages in PCK cells. After that, the process of cold-adaptation by growing the virus in PCK cells (20 passages) and eggs (18 passages) by gradually lowering the temperature of incubation from 28°C to 25°C was undertaken [28, 29]. At least 109 passages of A/Ann Arbor/6/60 were performed in different substrates. The resulting virus was shown to be cold-adapted, temperature-sensitive, and safe for mice.

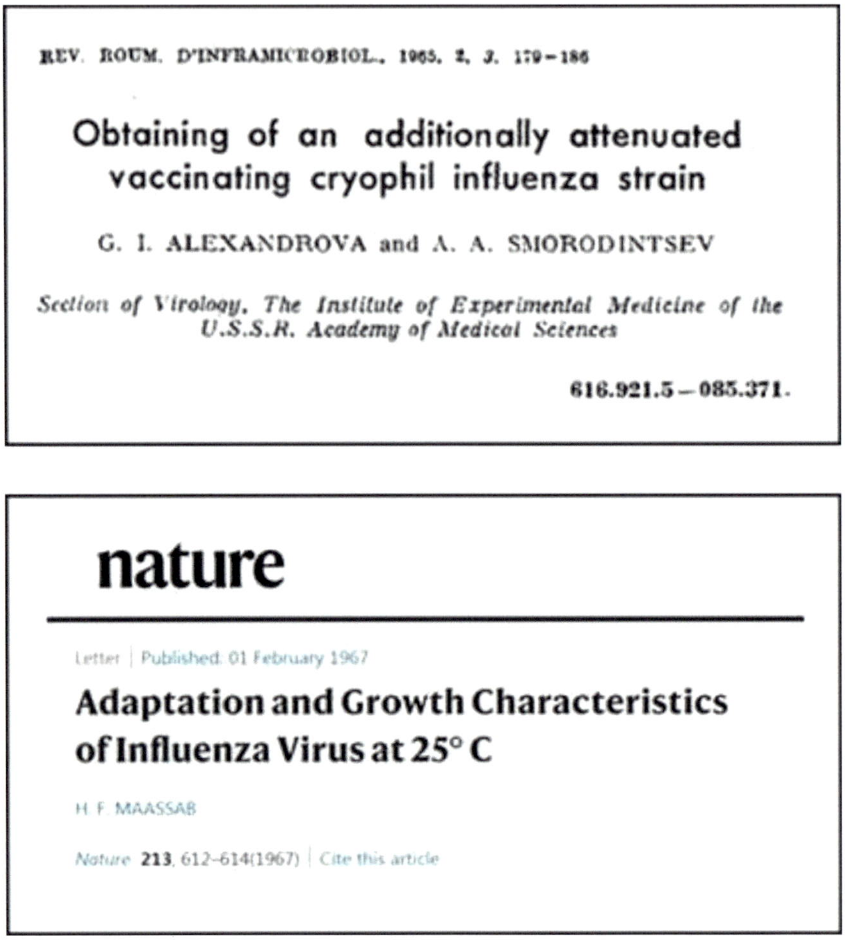

REV. ROUM. D'INFRAMICROBIOL., 1965, 2, 3, 179–186

Obtaining of an additionally attenuated vaccinating cryophil influenza strain

G. I. ALEXANDROVA and A. A. SMORODINTSEV

Section of Virology, The Institute of Experimental Medicine of the U.S.S.R. Academy of Medical Sciences

616.921.5–085.371.

nature

Letter | Published: 01 February 1967

Adaptation and Growth Characteristics of Influenza Virus at 25° C

H. F. MAASSAB

Nature **213**, 612–614(1967) | Cite this article

Fig. (4). The first publications on the development of cold-adapted live attenuated influenza vaccine. Top panel – screenshot of the title page of Alexandrova and Smorodintsev, 1965 [27]; Bottom panel – screenshot of the title page of Maassab, 1967 [28].

However, the passage history of this mouse-adapted line was not acceptable for use as a vaccine candidate. Thus, the selection of a new cold-adapted mutant of A/Ann Arbor/6/60 with a proven passage history was accomplished [29]. The original A/Ann Arbor/6/60 isolate was made to grow by gradually lowering the temperature of incubation, starting with 36°C and proceeding to 33, 30, 27, to 25°C [29, 30]. 30 passages in PCK cells and three passages in SPF eggs were performed. The resulting cold-adapted variant was later designated as A/Ann Arbor/6/60ca.

The method of attenuation by sequential passages in a new host is empirical due to the unpredictability of the genetic changes that the virus undergoes. The immunogenicity of passaged viruses is decreased faster than the occurrence of their attenuation. It is hard to predict the exact passage level where reactogenicity/safety and immunogenicity of the vaccine virus will be in a balance. In addition, repeated clinical trials on volunteers are necessary for each new passaged variant. Live attenuated reassortant vaccine of next-generation is free from these disadvantages.

In the early 1950s, a phenomenon of genetic reassortment of two competing viruses infecting the same cell was discovered [31, 32]. This observation gave rise to the era of obtaining reassortant viruses with specified properties, including influenza vaccine candidates [33].

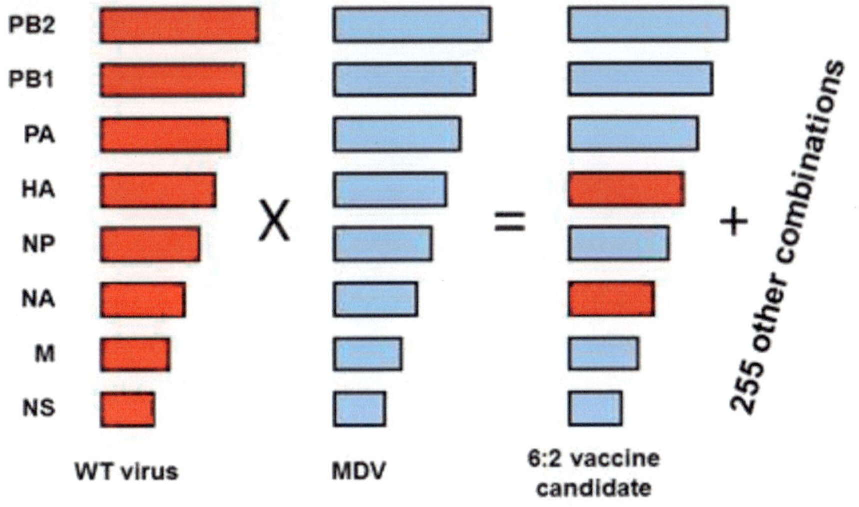

Fig. (5). The results of classical reassortment in eggs of WT influenza virus with master donor virus. As a result of co-infection, 256 genome combinations may occur, including the desired so-called 6:2 genome formula of the vaccine candidate. A vaccine candidate contains six gene segments encoding the internal proteins from the MDV (MDV, cold-adapted donor of attenuation or PR8, high yield donor of internal genes) and HA and NA genes from WT parental virus.

Reassortant LAIV candidates are generated by classical co-infection of the epidemiologically relevant WT parental strain with a cold-adapted master donor virus, as described elsewhere (Fig. **5**) [34]. The development of a 6:2 reassortant, by replacement of the genes coding for the internal proteins of WT virus with the corresponding genes of the MDV, is a reliable and reproducible method of attenuation. The surface glycoproteins HA and NA of the WT virus provide the targets for the immune response in the vaccinated host while the internal MDV proteins ensure that the reassortant virus is safe (LAIV) and highly grown (IIV). Cold-adapted master donor viruses repeatedly conferred an acceptable level of attenuation on each new emerging variant.

Two reassortant LAIVs developed in Russia and the US are now available commercially. The first one, licensed in 1987 for the prevention of influenza in persons 3 years and older as Ultravac (Microgen, Russia), is based on two cold-adapted MDVs, A/Leningrad/134/17/57 (H2N2) and B/USSR/60/69 [35 - 38]. The second, licensed in 2003 as FluMist (MedImmune, the US), is based on two other cold-adapted master donor viruses, A/Ann Arbor/6/60ca (H2N2) and B/Ann Arbor/1/66ca, and is used for the prevention of influenza in persons 2-49 years old [39 - 42].

The following Fig. (**6**) displays the most important milestones in the development of a live attenuated influenza vaccine.

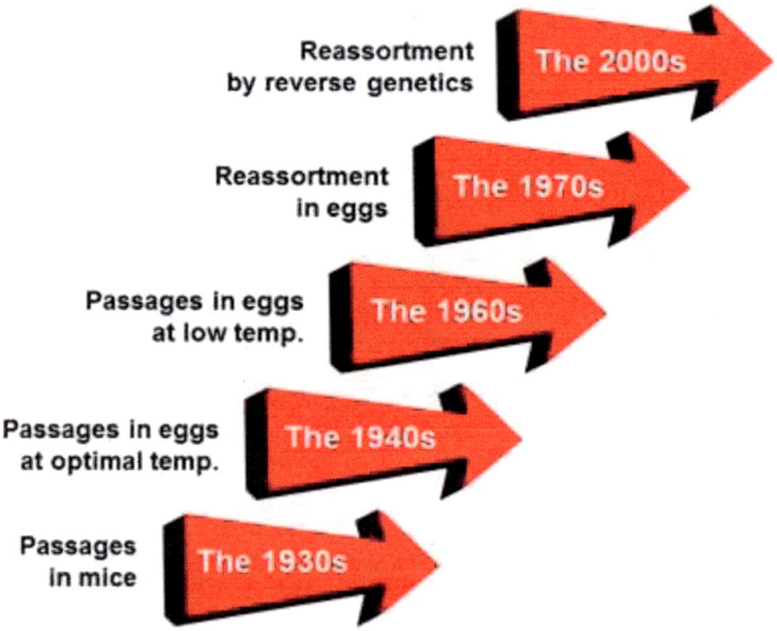

Fig. (6). Schematic evolution of live attenuated influenza vaccine.

Quadrivalent LAIV on the A/Ann Arbor/6/60ca and B/Ann Arbor/1/66ca MDVs' backbone is licensed in the US, Canada, and Europe. Trivalent LAIV on the A/Leningrad/134/17/57 and B/USSR/60/69 MDVs' backbone is licensed in Russia, India, and China [39, 43, 44]. LAIVs on the A/Leningrad/134/17/57 and B/USSR/60/69 MDVs' backbone are still entirely produced by classical reassortment in eggs. As for LAIV candidates based on A/Ann Arbor/6/60ca and B/Ann Arbor/1/66ca, they are generated by reverse genetics (Fig. **7**), with the same 6:2 genome composition as the classical vaccine (six genome segments from the attenuated, temperature-sensitive cold-adapted master donor virus type A or type B, and HA and NA from the seasonal circulating strain) [39, 43, 44].

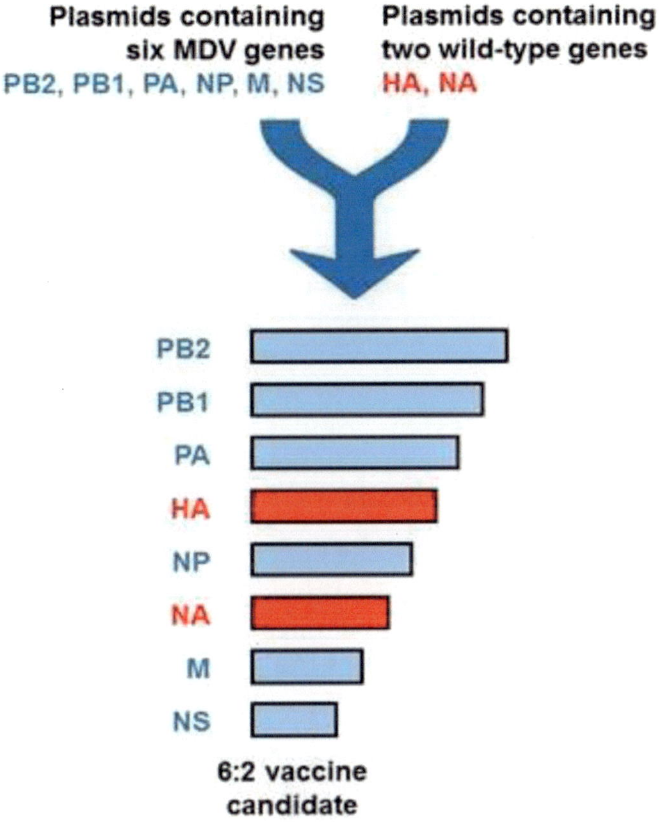

Fig. (7). Schematic diagram of LAIV vaccine candidate preparation by reverse genetics technique. Plasmids containing six MDV genes that control *ca/ts/att* phenotypes and two genes of WT virus that responsible for antigenic specificity are electroporated into Vero cells to generate the 6:2 vaccine strain.

Before 2009, LAIVs were successfully used to protect humans against seasonal influenza. In 2007, for the future pandemic preparedness, WHO brought together representatives from national immunization programs, regulatory authorities, vaccine manufacturers, and public health scientists to consider the state-of-the-art of LAIVs, and explore clinical and regulatory research to facilitate the potential use of these promising vaccines to control an epidemic and future pandemic influenza outbreaks [45, 46]. In 2009, the world was faced with the H1N1 influenza pandemic. New H1N1 pandemic LAIV candidates, thus, were generated and studied at the earliest [47, 48].

Along with the circulating pandemic H1N1pdm09 viruses, a number of highly pathogenic bird influenza viruses have already overcome the species barrier. As of

January 2020, in 17 countries of the world (Azerbaijan, Bangladesh, Cambodia, Canada, China, Djibouti, Egypt, Indonesia, Iraq, Lao People's Democratic Republic, Myanmar, Nepal, Nigeria, Pakistan, Thailand, Turkey, and Viet Nam), the total number of confirmed A (H5N1) bird flu cases among people reached 861, 455 of which were registered with the lethal outcomes [49].

The recent emergence of the H7N9 influenza virus in the human population in China resulted in global concern regarding its potential to become a pandemic [50]. On 31st March 2013, the public health authorities of China reported the first three cases of human infection with avian influenza A (H7N9) virus. New laboratory-confirmed cases and deaths in previously confirmed cases were reported daily. The patients did not appear to have a history of exposure to each other, but most of them appeared to have known exposure to poultry [51]. Genes of A (H7N9) viruses from human cases were most closely related to several potential parental strains of avian origin that circulated recently in China. Interestingly, both animal and human infections with H7N9 viruses have not been reported before the start of this outbreak.

If bird flu viruses acquire the ability of human-to-human transmission, it could result in a bird flu pandemic. This is confirmed by the fact that in the composition of the viruses' genome, triggering pandemics of XX-XXI centuries, there were always genes belonging to bird flu viruses [52].

Clinical Trial Plan for Pandemic Influenza Vaccines was developed at the international meeting on pandemic influenza vaccines (Bethesda, 2003), where a priority classification for pandemic vaccine candidates was proposed. The importance of this meeting may be attributed to the inclusion of LAIV into the global plan of WHO on influenza vaccines [46].

Inactivated vaccines including those that cannot multiply in the host cells. Inactivated vaccines are the most commonly used preparations for influenza prophylaxis. Licensed seasonal inactivated hemagglutinin-based influenza vaccines are of four types: (**i**) split virus, (**ii**) subunit, (**iii**) whole virus inactivated vaccines propagated in embryonated chicken eggs or cell-culture, (**iv**) recombinant hemagglutinin-based protein (Fig. **8**). Recently, numerous advances have been made in the development of IIV to replace inactivated whole-virus vaccines with split or subunit vaccines, which comprise less reactogenic alternatives [53].

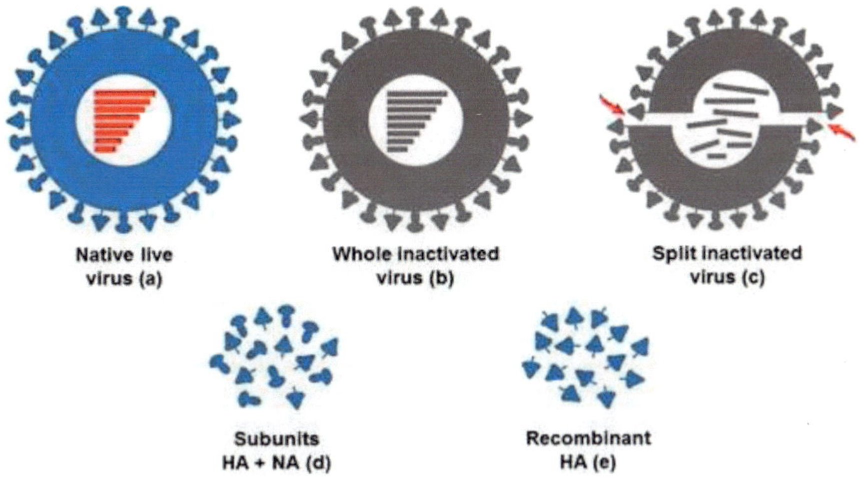

Fig. (8). Schematic diagram of licensed inactivated influenza vaccines. **(a)** Influenza virus virion structure and genome organization. **(b)** Whole inactivated vaccine. **(c)** Split inactivated vaccine. **(d)** Subunit vaccine. **(e)** Recombinant HA.

Typically, *whole-virus inactivated influenza vaccines* are produced in eggs, chemically inactivated with formaldehyde or BPL (β-propiolactone), concentrated and purified. In the past, candidate vaccine influenza A viruses for whole-virus IIV were developed using conventional reassortment technology. Interestingly, sometimes these reassortant viruses had a 5:3 gene constellation with three genes obtained from WT parental virus. For instance, in 2009, two candidate reassortant vaccine viruses (X-181 and X-181A) developed by New York Medical College, the US, using conventional reassortment technology, had the 5:3 gene constellation with three genes (PB1, HA and NA) obtained from A/California/07/2009 (H1N1) and the remaining five genes originating from A/PR/8/34 [54].

In *split virus vaccines*, the virus envelope undergoes disruption by diethyl ether or detergent treatment to expose all viral proteins [55]. *Subunit vaccines* involve additional purification steps to separate the nucleocapsid and lipids from the surface proteins HA and NA [56]. Split and subunit vaccines are more commonly used against seasonal influenza due to their reduced reactogenicity in comparison with the whole-virus vaccine [57]. Unlike hole virus-based vaccines, subunit influenza vaccines consist of purified viral HA or HA/NA proteins without the other viral components. These types of vaccines can be produced in eggs (*classical subunit vaccines*) or in baculovirus expression systems that carry the genes of the target antigen using novel manufacturing technologies (*recombinant subunit vaccines*) [55 - 61].

IIVs usually protect against antigenically close strains, but small antigenic changes are associated with a loss in protection.

Immune Response to Conventional Egg-Derived Influenza Vaccines

Inactivated influenza vaccines used in healthcare practice (whole virion, subunit, virosomal, split vaccines) are administered intramuscularly and are aimed at developing a protective humoral immune response (IgG) to the main antigens of the influenza virus HA and NA. An indicator of the potential efficacy of a vaccine is a protective titer of anti-HA antibodies of at least 1:40. Antibodies for NA cannot prevent infection but play an important role in protecting against severe forms of influenza as these prevent viral particles from escaping the infected cells, thus restraining the further spread of the virus in the body.

LAIV administered intranasally mimics natural infection and induces the same links of immunity as the causative agent of influenza. Along with the systemic humoral immune response (IgG), which is not so pronounced for LAIV, in response to LAIV replication in nasopharyngeal cells, local secretory IgA antibodies are induced and a mediated cellular immune response is stimulated [62, 63].

The completeness of the development of a post-vaccination immune response to infectious agents in humans depends on the ability of vaccine strains to induce precisely T- and B-cell immunological memory, which determines the body's protection during its subsequent contact with a "wild" virulent virus [64]. Cold-adapted LAIV is attenuated and is unable to cause infection in the lower respiratory tract due to the higher temperature in these tissues [39]. One of the advantages of LAIV is its ability to induce cross-protection against drift influenza virus variants.

The response to vaccination with a live or inactivated vaccine depends in part on the immune status of the person being vaccinated, which is the result of a history of previous exposure to the influenza virus. In particular, children who have not previously had influenza virus exposure have a better immune response to LAIV than adults. In seronegative children, serum IgA and IgG antibodies to HA, in response to LAIV immunization, exhibited higher indices than those for IIV [65]. The low level of serum antibodies in children and seronegative adults, produced in response to IIV administration, necessitates two-fold immunization, whereas one dose is sufficient for seropositive individuals [66].

An inverse relationship is observed in vaccinated adults who have had previous encounters with the influenza virus; they have an inactivated vaccine that is more immunogenic.

Many of the differences between the response to LAIV or IIV vaccination in adults and children can be explained as follows. Since LAIV replication in the respiratory tract may be partially limited in adults due to preexisting immunity to the related seasonal influenza virus, the live vaccine replicates at higher titers and over a longer period of time in seronegative *versus* seropositive individuals [67, 68]. For this reason, LAIV is a much greater antigenic stimulus for children than for adults. Conversely, in individuals with a preexisting experience of a related antigen, the antibody response to IIV is higher since this immune response is a secondary booster. This explanation logically follows the advantage of LAIV as a pandemic vaccine, when a primary immune response is generated in response to vaccination as well as to disease [69].

The use of IIVs as seasonal vaccines is justified for seropositive adults and those suffering from cardiopulmonary pathology, who, in response to intramuscular administration of the vaccine, develop serum humoral antibodies to directly injected strains of the virus.

Immune responses to natural infection with influenza virus in humans are relatively broad and long-lived, but influenza viruses can escape from these responses over time owing to their high mutation rates and antigenic flexibility. Vaccines are the best available countermeasure against infection, but influenza vaccine effectiveness is lower compared with the WT virus. Furthermore, inactivated influenza virus vaccines focus on the induction of systemic IgG responses but do not effectively induce mucosal IgA responses. A better understanding of how natural infection induces broad and long-lived immune responses will be key to developing next-generation influenza virus vaccines [64].

Cell Culture-Derived Influenza Vaccines

Mutations occur in HA as human influenza viruses adapt to grow in eggs [70, 71] (Fig. **9**). This can dramatically alter the antigenicity of the virus. Most attempts to reselect an influenza virus possessing sequence changes in HA characteristics of mammalian cells by multiple passages in MDCK cells of the egg-derived isolate were not successful [72]. In contrast, Kuwahara *et al.* [73] demonstrated that by serial egg passages of a cell-propagated virus, they were able to successfully isolate an egg-adapted influenza A (H3N2) virus without amino acid substitutions at the antigenic sites of HA but with multiple substitutions in NA. Antigenic analysis of this egg-adapted virus indicated that its antigenicity did not differ from that of the WHO prototype cell-propagated vaccine virus. This strategy may facilitate egg-based vaccine production without antigenic alterations in HA by egg adaptation.

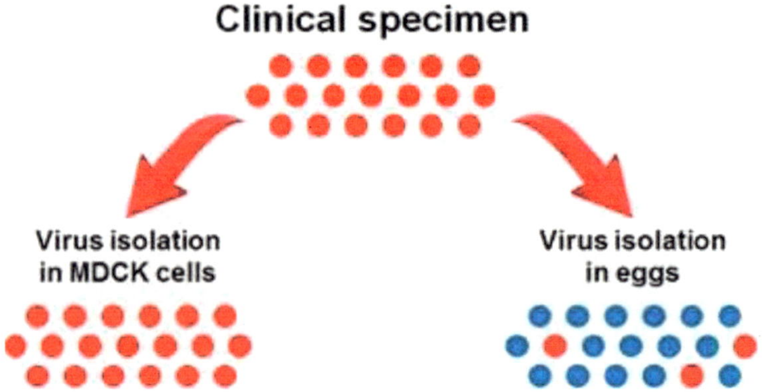

Fig. (9). Selection of influenza viruses derived from clinical specimens on MDCK cells and in eggs. Red circle – virus particle identical to particles from a clinical specimen. Blue circle – mutated virus particle that appeared after replication in eggs.

In the 1980s, it was found that H3N2 viruses were not fully isolated in eggs than in cell culture [74]. In the next years, the situation became even more complicated. Since the late 1990s, influenza vaccine manufacturers have been faced with a serious problem. Surprisingly, human H3N2 influenza viruses were not isolated in eggs and it was hard to select an egg-derived strain for vaccine production. The reason could be that human H3N2 viruses isolated in cell culture were reported to bind with a high affinity to α2-6-linked sialosides, while viruses isolated in eggs often had increased specificity for α2-3-linked sialosides [75 - 77].

Replicative advantages in tissue culture of egg-adapted influenza virus over tissue culture-derived virus for vaccine production were found in the 1990s [20]. It was suggested that "…vaccine manufacturers who are investigating the use of tissue culture for vaccine production should minimize passage levels of egg-derived virus and beware of the displacement of the original virus with variants similar to those derived in eggs, which are often antigenically distinct…" [20]. However, WHO made recommendations on the composition of influenza virus vaccines based on egg-isolated reference viruses until 2018, and only on 27 September 2018, the first recommendations on non-egg-based vaccines were made for their use in the 2019 southern hemisphere influenza season [78].

Production technologies culminating in cell-based influenza vaccines have been in development since the 1930s [16] but became a relevant technology only in the 2000s [79, 80].

Compared to egg-based technology, cell-culture-derived influenza vaccines reduce vaccine production time and risk of contamination during production.

They are safe for those with an allergy to eggs; also, animal-component-free production is feasible. As mentioned before, egg passaging might induce adaptive changes for growth in eggs. Cell-derived vaccine is devoid of these disadvantages. Thus, in the future, cell culture-derived vaccines may replace the conventional time-consuming, labor-intensive egg-based influenza vaccine production method.

Two approaches can be utilized for the development of cell-based vaccines: (**i**) some steps of vaccine preparation are carried out in eggs and final steps are performed in cells or (**ii**) all steps of vaccine preparation are performed in cells (Fig. **10**).

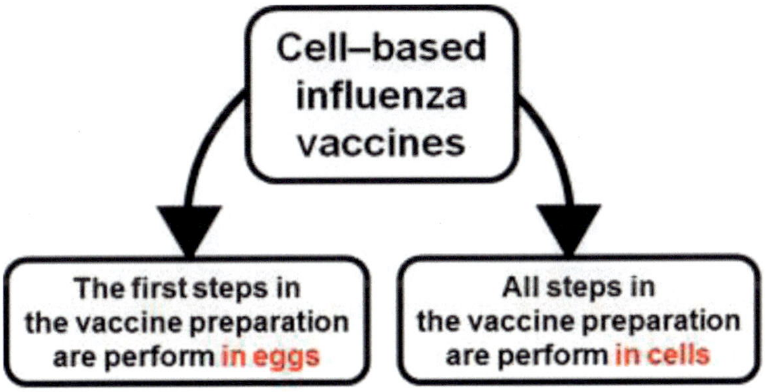

Fig. (10). Two main approaches to the development of cell-based influenza vaccines.

(**i**). Vaccine candidates and master donor viruses for reassortant LAIV and IIV are grown well in cell culture without additional adaptation [81 - 83]. This property can be used in production to prepare cell-derived vaccine batches. In 2011, Russian scientists published the results of the development of live attenuated cell-based pandemic influenza vaccine, Vector-Flu [82]. In 2012, United States Food and Drug Administration (FDA) announced the approval of the first seasonal influenza cell-based subunit inactivated influenza vaccine, Flucelvax [84]. Interestingly, in both the cases, the manufacturing process did not involve eggs; however, the vaccine could not be considered as completely egg-free because the vaccine seed virus supplied by WHO was passaged in eggs [84] or reassortant vaccine candidate was derived by co-infection of two egg-derived parental viruses in eggs [82, 85]. In fact, these are cell-amplified preparations only.

This approach possesses one serious disadvantage; vaccine viruses contain egg-adapted mutations which may alter the antigenicity of the virus. Problems with unwanted egg-adapted mutations and difficulties of reassortment can be solved by reverse genetics approaches.

(**ii**). The entire process of vaccine development, including WT virus isolation, could be carried out in tissue culture [86]. This is a more logical approach because, in this case, vaccine candidates will be free of egg-adapted mutations. The use of tissue culture as a substrate could make influenza vaccine manufacturing independent of the global egg supply and enable easy scaling up of the production process. Palker *et al.* [86] compared the protective efficacy of an egg-derived reassortant LAIV candidate based on egg-isolated A/New Caledonia/20/99 (H1N1) virus with MDCK-derived reassortant LAIV candidate based on MDCK-isolated A/New Caledonia/20/99 (H1N1) virus. The study conferred similar levels of protection of ferrets challenged with WT parental virus.

However, reassortment in tissue culture encounters unpredictable difficulties in the regular development of 6:2 reassortant candidates. For unknown reasons, reassortants generated in MDCK cells usually possess the majority of genes of the WT parent *(I. Kiseleva, pers. comm.)*.

Another vaccine, so-called ΔNS1, can be conditionally attributed to live attenuated cell-based vaccines [87 - 89]. Its attenuation was attained using a novel attenuation approach - the deletion of the NS1 gene encoding the multifunctional non-structural protein, which counteracts the interferon-mediated antiviral response. ΔNS1 viruses are replication-deficient in interferon-competent systems and are unable to overcome the cellular antiviral defense.

Synthetic Influenza Vaccines

Three approaches to the development of licensed commercial influenza vaccines are currently in use. They include (**i**) live vaccines, (**ii**) inactivated (whole virion, split and subunit) vaccines, and (**iii**) recombinant HA.

Conventional seasonal influenza vaccines elicit HA head-specific neutralizing antibody responses. However, the HA head domain is highly flexible, and mutations can be easily introduced, resulting in antigenic drift. Since vaccine production takes several months to generate a sufficient number of doses, the circulating virus may differ from the vaccine strain. Obviously, conventional seasonal influenza vaccines cannot protect against unexpected emerging influenza viruses. Furthermore, the majority of licensed vaccines are produced in eggs, which sometimes leads to egg-adapted mutations that may significantly lower vaccine effectiveness. In addition, current influenza vaccines are often not strongly immunogenic and elicit only low antibody responses [64, 90 - 92].

Many of these problems can be avoided by the development of alternative approaches to these conventional influenza vaccines. Several novel technologies that may improve the vaccine production process have been described recently. These disruptive technologies could bring dramatic changes in the current system of influenza vaccine production by the development of long-lasting, broadly cross-reactive, and protective universal influenza virus vaccines [90]. Major efforts to improve approaches to vaccine development are underway.

Here, we have modified Fig. (**2**) (see section "Substrates for influenza vaccine development") to more detailed Fig. (**11**). As can be seen from Fig. (**11**), the development of synthetic influenza vaccines is the third platform for influenza vaccine production. Gene synthesis technique combined with reverse genetics approaches allows the generation of IIV and LAIV seed viruses of desired properties within short time frames.

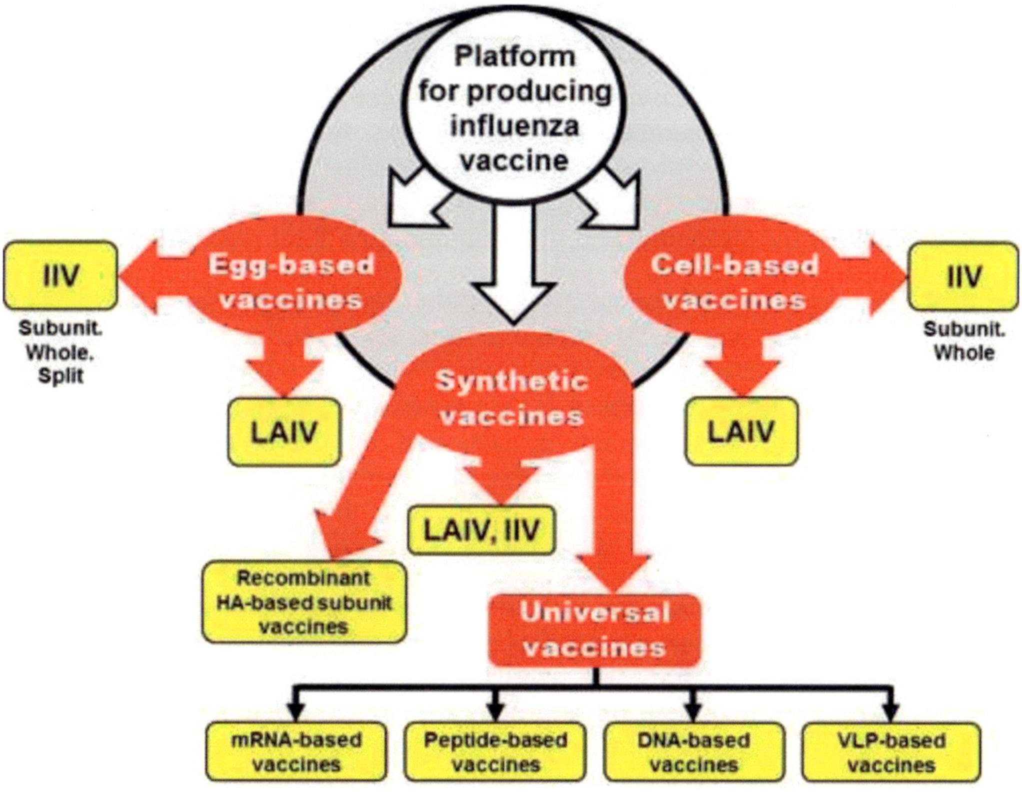

Fig. (11). Conventional and novel platforms for influenza vaccine production [93].

Synthetic LAIV candidates are 6:2 reassortant viruses that have been generated using reverse genetics techniques. With the development of plasmid-based reverse genetics techniques, it is now possible to engineer recombinant influenza viruses entirely from full-length complementary DNA copies of the viral genome by transfection of vaccine-approved susceptible cells (Fig. 7). LAIV candidates for FluMist (the US) are currently produced by reverse genetics and amplified in eggs. In contrast, Russian LAIV candidates are still under development by classical reassortment in eggs. One of the reasons why the classical reassortment technique for LAIV production is preferred over reverse genetics approaches is because it allows for the natural selection of a variant with a gene constellation that improves growth properties of the egg [61, 94].

Synthetic IIV candidates like synthetic LAIV candidates are also reassortant viruses that have been generated using reverse genetics techniques [95]. We should keep in mind that conventional reassortment does not always result in the formation of reassortants of the desired gene constellation. Sometimes 6:2 genome formula cannot be obtained [96, 97]. Then reverse genetics techniques come to the rescue. The advantage of the reverse genetics method is that it is usually possible to generate a reassortant of the desired 6:2 gene constellation. Perhaps there are some objective reasons why conventional reassortment does not work. Possibly the 6:2 constellation is not optimal for a given pair of parental viruses. In this case, the forced construction of the 6:2 reassortant using the reverse genetics method can lead to the creation of a virus that has properties that are not optimal for the vaccine strain, for example, low viral replication.

Recombinant hemagglutinin-based protein technology has been briefly described above (see section "Inactivated vaccines including those that cannot multiply in the host cells"). Many problems associated with the use of chicken embryos, the low affinity of current H3N2 viruses to eggs, and the necessity to speed up the vaccine development process can be solved using recombinant subunit vaccines. On the other hand, the low immunogenicity of this type of vaccine is its drawback, as well as of conventional subunit vaccines. The need for repeated vaccination and the use of adjuvants are obvious.

Universal influenza vaccines can induce a cross-protective broadly neutralizing immune response against conserved viral antigens that may overcome the above-mentioned drawbacks of currently used vaccines. A truly universal influenza vaccine would offer a combination of protection from antigenic drift and shift and ideally confer lifelong immunity. The development of universal influenza vaccines will help to reduce the severity of influenza epidemics and pandemics.

Over 20 different approaches are described as strategies for the development of universal influenza vaccine, such as improvement of hemagglutinin-targeting vaccines (chimeric hemagglutinins, mosaic hemagglutinins, headless hemagglutinins, mosaic nanoparticles); M2-targeting vaccines; internal protein-targeting vaccines; neuraminidase-targeting vaccines; technologies to improve immunogenicity and longevity of immune responses. These approaches include recombinant, protein-based and virus-like particle vaccines, nucleic acid-based vaccines, recombinant bacterial vector-based vaccines, recombinant viral vector-based vaccines, the M2 ectodomain-based vaccines, HA stalk domain-based vaccines, NA-stalk domain-based vaccines, *etc.* [53, 58, 59, 90, 98 - 111].

In the review of Jazayeri and Poh [98], all these numerous approaches are combined in four major platforms: DNA-based vaccines, RNA-based vaccines, peptide-based vaccines, VLP-based vaccines [109]. In the review of Rajão and Pérez [58], the approaches to the development of universal influenza vaccines are also combined in four major platforms but slight differently: **(i)** viral vector vaccine, **(ii)** nucleic acid-based, **(iii)** recombinant, protein-based and **(iv)** virus-like particle vaccines. We will stick to the first classification (Fig. **11**) [98].

An ideal universal influenza vaccine is effective against all influenza A and B viruses subtypes/lineages regardless of any mutations in HA and/or NA. The objective of this vaccine is to induce cross-protective and long-lasting immune responses by stimulation of both arms of the immune system. To design this type of vaccine, the highly conserved epitopes present in HA, NA, or M2 and internal proteins such as NP and M1 should be targeted to induce cross-protective antibodies and T-cells. The development of a new universal influenza vaccine can be significantly facilitated by newly-developed platform technologies such as multi-epitope, VLP, DNA, and mRNA-based vaccines.

DNA-based vaccines are created by incorporating sequences coding for influenza proteins into a DNA bacterial plasmid [59]. The advantage of DNA vaccines is that the production process starts with the nucleotide sequence of HA, which is artificially synthesized and introduced into a productive context that is completely independent of the whole virus [60, 98, 102, 112 - 116]. Influenza DNA vaccines can induce broad-spectrum protective immune responses in various animal models. The DNA-plasmid vaccine technology provides a rapid, highly scalable, and safe alternative to conventional influenza vaccine production without the need for eggs [58]. However, the levels of cell-mediated and humoral responses induced by the introduction of a DNA vaccine are often insufficient for the development of immunity against pathogens and are typically used with adjuvants to enhance immunogenicity.

Another nucleic acid-based vaccine platform, **RNA-based technology,** is even more attractive since it harbors only components required for protein expression, is rapidly degraded, and does not interact with the host genome [58]. Thus, messenger RNA (mRNA)-based vaccines offer safety advantages in comparison to DNA-based vaccines. RNA-based vaccines promise quick, strain-independent production with high yields from a small manufacturing footprint. Two types of RNA-based influenza vaccines are currently being developed: self-amplifying [117 - 119] and non-replicating mRNA vaccines [118].

Modified mRNA vaccines encoding HA of potentially pandemic avian influenza viruses formulated with lipid nanoparticle generated robust immune responses in several animal models, such as these protected them from lethal challenge infection [120], and were found to be safe and immunogenic in humans in a randomized, double-blinded, placebo-controlled phase I clinical trial [120].

Peptide-based vaccines. Efforts towards peptide-based universal vaccines have involved highly conserved single or multi-epitope small peptides. The design of these vaccines is based on the reduction of disease severity rather than the prevention of infection [53]. When there are no protective antibody responses to newly evolved influenza viruses, T-cells could provide heterosubtypic immunity to various influenza virus subtypes and even unrelated viruses because of conserved peptide homology [121, 122]. Several vaccine candidates based on an *E. coli*-expressed fusion peptide containing different epitopes and composed of single or multiple B- or T-cell epitopes are in development or have been tested in clinical trials and found to be safe, inducing cellular immune response [123 - 126].

Virus-like particles (VLPs) are non-infectious immunologic particles that closely mimic the structure of live virus particles by expressing the desired viral antigens and retaining natural structural conformation. VLPs are obtained from the self-assembly of viral structural proteins carrying viral antigens. VLPs are produced by co-expression of influenza virus structural proteins in mammalian cells, insect cells, or plants without a viral genome, representing a high safety feature [98, 127].

As the alternative to the commercial influenza vaccines which are still based on the eggs, advances in cell culture technologies involving plant, insect, and mammalian cells could provide egg-independent manufacturing options to produce VLP vaccines [61] especially pandemic VLP vaccines [98]. The performance of an insect cell-derived influenza pandemic H1N1pdm09 VLP vaccine was successfully demonstrated in clinical trials on adult volunteers and showed the persistence of antibodies up to 24 months after vaccination [128].

Universal vaccines are presently under development in the laboratories worldwide. Recent scientific progress in the development of the potential next-generation universal influenza vaccines is tremendous.

Cross-Protection Potency of Influenza Vaccines

Conventional IIVs primarily induce virus-specific adaptive antibody responses, *i.e.* these are strongly strain-specific and poorly effective against mismatched viruses [58, 129] (Fig. **12**). IIV mostly induces antibodies against the globular head domain of the HA protein. HA-head-specific antibodies are only protective against closely antigenically related viruses. Only a limited number of studies describe possibilities to increase the level of cross-protection of IIV. Ni *et al.* [130] suggested treating viral antigen at low pH. According to the authors, such treatment is associated with increased cross-protection against heterologous viruses. Inactivated vaccines treated with low pH could be particularly beneficial in cases of mismatch of vaccine and circulating virus.

In contrast to traditional inactivated vaccines, LAIV is capable of inducing broad-spectrum and long-lasting immune responses (Fig. **12**), making them an attractive option for pandemic preparedness, especially in low-income countries with very high population density. WHO considers the expansion of LAIV production as a promising strategy to increase the influenza vaccine supply in the pandemic situation [131].

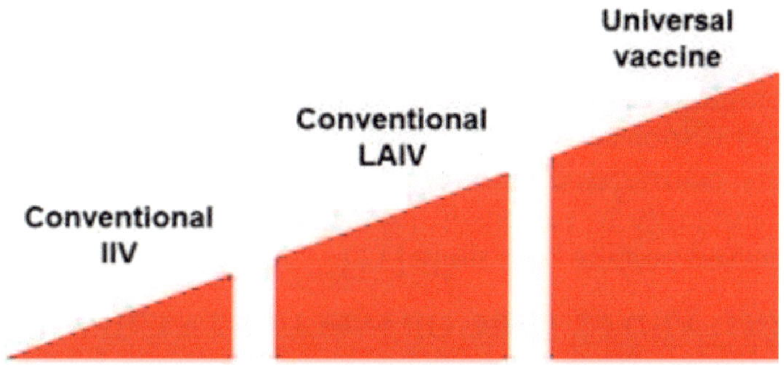

Fig. (12). Cross-protective efficacy of different types of influenza vaccines [93].

LAIV is administered as a nasal spray and requires replication of vaccine virus in the mucosa of the upper airways to induce protection. By mimicking natural infection, LAIV induces both humoral and broad T-cellular immune responses [132 - 134]. Importantly, LAIV may elicit not only homologous but heterosubtypic protection. The possibility of heterosubtypic protection by LAIV has been proven by numerous studies.

Mono- and trivalent LAIVs were shown to be cross-protective against genetically different influenza B/Victoria and B/Yamagata viruses based on virological and histological data and clinical signs in ferrets [135, 136]. LAIV protects animals challenged with heterosubtypic influenza strains [137 - 139] through induction of tissue-resident memory T-cells in the lungs [140]. More importantly, human studies have found that the LAIV has shown the potential to confer broader protection than the IIVs in children when used in school settings [141, 142]. In contrast to IIV, LAIV induces cross-protective $CD8^+$ T-cells in young children, and these T-cells may provide relevant protection in the case of a new pandemic [44, 143].

The most attractive and promising influenza vaccines to be widely cross-protective are the universal vaccines that are designed to induce cross-protective, broadly neutralizing immunity against conserved epitopes of the influenza virus (Fig. **12**). The broad elicitation of cross-protective and neutralizing antibodies by the universal vaccine depends on stimulating both the humoral and cell-mediated immune responses. It has been shown that vaccines based on the highly conserved epitopes of HA and NA induce a cross-protective T-cell response [53, 58, 90, 129, 144]. Thus, the problems related to the high variability of surface proteins of the influenza virus that lead to the need for annual updating of seasonal vaccines may be overcome [53, 58].

In addition to HA and NA humoral immune responses, some other conserved influenza viral antigens could be recognized by the CD8 T-cells and elicit cross-reactive antibodies – an extracellular domain of the M2 protein, internal proteins, such as NP, PA, PB1 and M1 [145]. Vaccines based on these antigens are able to induce broad immune responses [146]. If influenza vaccination results in low levels of antibody responses to newly evolved influenza viruses, these antibodies can have broadly neutralizing activity [58, 90, 129], and T-cells could provide heterosubtypic immunity to various influenza virus subtypes or unrelated viruses because of conserved peptide homology [121, 122].

PAST, PRESENT, AND FUTURE OF INFLUENZA CHEMOTHERAPY

All antivirals, including anti-influenza drugs, could be classified into two main categories - direct-acting antivirals (DAAs) and indirectly acting drugs (Fig. **13**). Antivirals could (**i**) target different viral proteins or stages of the viral life cycle (DAAs) or (**ii**) act indirectly as symptomatic and etiotropic drugs (for instance, pain relievers, fever reducers, vitamins, *etc.*) or through a change of the metabolism of the host cell which was modified under the influence of a viral infection (drugs affecting lipid, copper, or calcium metabolism).

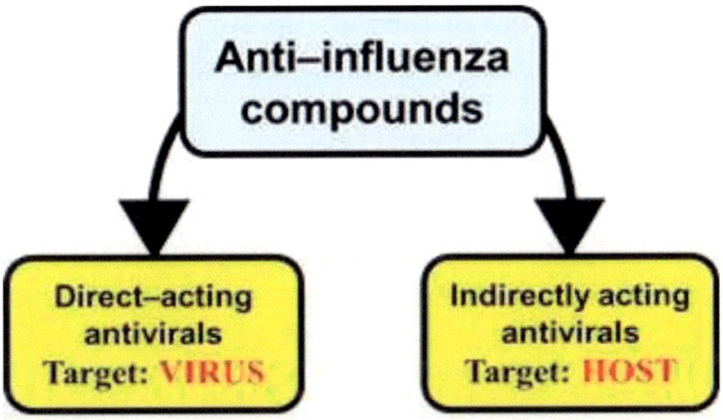

Fig. (13). Target-dependent types of anti-influenza virus compounds.

All currently licensed influenza drugs belong to the group of DAAs [147]. Compared to indirectly acting host-directed experimental antivirals, DAAs have a lower tendency for undesirable side effects. However, the rapid development of viral resistance has emerged as the predominant liability of a decrease in efficiency of some DAAs, especially when directed against RNA viruses with error-prone polymerases, such as influenza viruses [11]. Drug resistance will be discussed in Chapter **5**.

There are only four FDA-approved influenza antivirals recommended by CDC for use in 2020-2021 influenza season against recently circulating influenza viruses - Rapivab® (peramivir), Relenza® (zanamivir), Tamiflu® (oseltamivir phosphate, also available as generic) and Xofluza® (baloxavir marboxil) (Table **1**) [8, 148 - 151]. These antivirals belong to two classes - neuraminidase blockers interfering with the release of progeny virus particles [8, 152 - 154] and cap-dependent endonuclease inhibitors of the influenza virus polymerase PA subunit [10, 149, 154].

Direct-Acting Antivirals

Adamantanes. In the 1960s, William W. Prichard discovered a new class of antiviral agents, adamantanes, and patented the chemical formula of alpha-methi--1-adamantylmethyl-amin hydrochloride (rimantadine hydrochloride) [155] and the method of its synthesis [156]. Rimantadine was shown to be protective against influenza A infection in a different animal model [157] and in humans [158].

In the 1970s, Latvian pharmacologist, Janis Polis (1938-2011) discovered an innovative method for producing alpha-methil-1-adamantylmethyl-amin hydrochloride [159]. His method of synthesis of rimantadine was fast and cheap. For his invention and introduction of the antiviral preparation rimantadine in the

USSR, he was called "the father of the anti-flu bomb." He has helped thousands of people. Unfortunately, no one could help him solve his health problems. 72-year-old Polis, who suffered from serious leg vein problems, ended his life in oblivion and poverty in a nursing home [160].

Adamantanes (amantadine and rimantadine) have direct antiviral effects against influenza A viruses. They were the first antivirals approved for the treatment of influenza A virus infections. Amantadine was approved for clinical use in 1966, followed by rimantadine in 1993 [11]. Both antiviral compounds were found to be highly effective for the treatment and prevention of influenza A infection with an efficacy rate of up to 90% [161, 162]. Adamantanes target the viral M2 ion channel membrane protein, preventing dissociation of the viral RNP genome from the matrix protein by blocking M2-mediated diffusion of protons into virions located in maturing endosomes, resulting in blocking virus entry into the cells [163, 164].

Table 1. Antivirals for treatment and chemoprophylaxis of influenza.

Active Ingredient	Trade Name	Mechanism of Antiviral Action	Route of Administration	Target Virus Type
Rimantadine	Flumadine®	M2 channel blocker	Oral	A
Oseltamivir[1]	Tamiflu®	NA inhibitor	Oral	A and B
Zanamivir[1]	Relenza®	NA inhibitor	Inhaled	A and B
Peramivir[1]	Rapivab®	NA inhibitor	Intravenous	A and B
Laninamivir	Inavir®	NA inhibitor	Inhaled	A and B
Baloxavir[1]	Xofluza®	Polymerase inhibitor	Oral	A and B
Favipiravir	Avigan®	Polymerase inhibitor	Oral	A and B
Pimodivir	No name	Polymerase inhibitor	Oral	A
Umifenovir	Arbidol®	HA inhibitor	Oral	A and B

[1]Recommended for treatment and chemoprophylaxis of influenza in the 2020-2021 influenza season by CDC [8, 148].

Adamantanes have two major disadvantages: **(i)** no specific activity against influenza B viruses; **(ii)** development of resistance of influenza A viruses to adamantanes. Unfortunately, the limited efficacy of amantadine and rimantadine against recent influenza virus strains spurred the development of new therapeutic drugs and approaches.

For many years, FDA-approved amantadine (generic) and rimantadine (Flumadine® and generic), fusion inhibitors, which impair the virus entry, were the gold standard of therapy for the treatment and chemoprophylaxis of influenza

A virus infection. However, today, many strains of influenza virus are resistant to these drugs, and CDC has not recommended the use of amantadine and rimantadine for recently circulating influenza viruses [8].

Although rimantadine has been excluded from the recommended antivirals for influenza treatment, it has an additional beneficial effect, *i.e.* it possesses a wide spectrum of antitoxic activity. It is well known that primary viral pneumonia is the most severe pulmonary manifestation of influenza infection both in humans and in mice inoculated with a mouse-adapted virus [165 - 171]. Viral pneumonia develops on days 6-8 after intranasal infection in mice. Depending on the dose, the mortality rate may reach 100%. Conversely, non-mouse-adapted influenza viruses were shown to kill mice by a so-called toxic effect when massive doses were administered [172, 173]. Acute pulmonary edema develops on days 2-4 after intranasal or intravenous inoculation of mice.

We modeled acute toxicity (development of acute pulmonary edema) of non-mouse-adapted influenza viruses or Newcastle disease virus in outbred white mice by intravenous (in *sinus orbitalis*) or intranasal administration of the massive dose of the virus ($\geq$ 8.0-8.5 $\log_{10}$ EID_{50}) *(I. Kiseleva, pers. comm.)*. Lethality from acute pulmonary edema was monitored daily for the first three days post-infection. The massive dose of tested viruses caused 70-90% of mortality during the first three days post-infection in the control group of mice. Pathological examination of the lungs of dead mice showed acute pulmonary edema, affecting all lobes of the lung.

At the time of the toxic death, the lungs of control mice inoculated with viruses were hemorrhagic. In contrast, if mice were pretreated with rimantadine, they survived after the inoculation of the virus. Preventive scheme of rimantadine administration to mice (-24 hrs. and -1 hr. before virus) exhibited protection as indicated by the increase in the number of survived mice (Table **2**). It is noteworthy that *in vivo,* the influenza B strains were also susceptible to rimantadine.

Interestingly, in 1965, Kenneth W. Cochran and co-authors demonstrated a similar protective effect of amantadine hydrochloride against B/Lee/40 influenza virus [174]. Unfortunately, the authors did not specify a dose of the virus they used. We can assume that the dose was very high and the non-specific antitoxic effect of amantadine hydrochloride was first discovered.

Acute pulmonary edema was also caused by intranasal administration of exotoxin of *Staphylococcus aureus* (*anatoxinum staphylococcus*) or intravenous (in *sinus orbitalis*) administration of adrenaline or chloramine to mice. It is noteworthy that rimantadine was protective against acute generalized pulmonary edema caused by chloramine (*Chloraminum B*) and exotoxin of *S. aureus*. In contrast, acute pulmonary edema caused by intravenous inoculation of adrenaline (epinephrine) was not found to be prevented by pretreatment with rimantadine (Table **2**). Probably, a different mechanism exists for the development of acute pulmonary edema by adrenaline.

These results suggest that rimantadine might be used for the prevention of the toxic effects of respiratory viruses even if they are rimantadine-resistant.

Besides M2 channel protein blockers, other proteins that are involved in viral replication are known as targets for the direct-acting antivirals - neuraminidase, polymerase subunits PB2, PB1, and PA, and hemagglutinin.

Table 2. Antitoxic activity of rimantadine in mice *(I. Kiseleva, pers. comm.)*.

Acute Pulmonary Edema Caused by the Administration	Route of Administration	Lethality, n (%)	
		Control	Treated[1]
A/Singapore/1/57 (H2N2)	In *sinus orbitalis*	36/39 (92.3)	8/40 (20.0)
B/Lee/40	In *sinus orbitalis*	25/31 (80.7)	7/28 (25.0)
B/Hong Kong/5/72	Intranasally	24/30 (80.0)	8/30 (26.7)
B/Khabarovsk/85/79	Intranasally	23/30 (76.7)	7/27 (25.9)
Newcastle disease virus	Intranasally	41/55 (74.6)	18/62 (29.0
Exotoxin of *Staphylococcus aureus*	Intranasally	33/40 (82.5)	8/40 (20.0)
Adrenaline	In *sinus orbitalis*	29/40 (72.5)	31/40 (77.5)
Chloramine	In *sinus orbitalis*	54/60 (90.0)	14/60 (23.3)

[1]Rimantadine was given orally in a dose of 150 mg/kg/day for two days, starting the day before toxic agent exposure (-24 hrs. and -1 hr.). Control animals received an equal volume of phosphate-buffered saline.

Neuraminidase inhibitors are chemically related antivirals that block the viral neuraminidase enzyme and exhibit activity against both influenza A and B viruses. Four representative drugs from the class of NA inhibitors are oseltamivir, zanamivir, peramivir, and laninamivir [9].

The first three NA inhibitors are FDA-approved and recommended for treatment and chemoprophylaxis of influenza by CDC [8, 147, 175]. The most widely utilized NA inhibitor is oral oseltamivir which is used in many countries for the treatment of suspected or proven influenza [175 - 177]. Zanamivir is the inhibitor of influenza A and B neuraminidases, including some variants resistant to other NA blockers [178]. Zanamivir is used in powder form and inhaled through the mouth *via* an inhaler. The third common NA inhibitor, peramivir, is given by intravenous infusion [150, 151, 154].

Single administration to mice and ferrets of another long-acting NA blocker – laninamivir – was shown to be more effective than compared with that of repeated administrations of oseltamivir or zanamivir [179]. Clinical trials of laninamivir inhaled as a suspension or as dry powder were completed in different age groups [180, 181]. This NA inhibitor is used for the treatment and prophylaxis of influenza virus A and B infection. The clinical effectiveness of laninamivir was confirmed by Japanese researchers, and it has been approved for the treatment and prophylaxis of influenza in Japan [175, 180]. Laninamivir has been under clinical evaluations in other countries [182].

The usefulness of neuraminidase inhibitors is limited by the need to administer them within 48 hours of symptoms developing.

Polymerase inhibitors. The influenza virus RNA-dependent RNA polymerase consists of three essential viral replication subunits – PB2, PB1, and PA (for more details, see Chapter **2**). Several antivirals specifically target one or another subunit of the influenza polymerase complex.

PB2 inhibitor pimodivir possessing selective activity against influenza A viruses [183], has shown virologic efficacy in patients with uncomplicated influenza and hospitalized patients, and is advancing in clinical development [184].

PB1 inhibitor favipiravir selectively inhibits the RNA-dependent RNA polymerase of RNA viruses, including influenza viruses [185, 186]. The mechanism of the antiviral activity of favipiravir is not clear. Some believed that it could be misincorporated in a viral RNA or could act by binding to conserved polymerase domains [185 - 187]. Favipiravir was approved in Japan for the treatment of influenza viruses resistant to other agents; however, in randomized, placebo-controlled clinical trials, favipiravir has shown inconsistent clinical effects in uncomplicated influenza [10, 188].

Another PB1 inhibitor, ribavirin, has not demonstrated convincing clinical efficacy in influenza treatment [189].

Baloxavir marboxil is an oral prodrug that is rapidly converted to its active form baloxavir acid and selectively inhibits the cap-dependent endonuclease activity of the PA subunit required for viral mRNA transcription [190]. *In vitro*, baloxavir acid inhibits replication of seasonal human and avian influenza viruses, including strains resistant to NA inhibitors and adamantanes, at very low concentrations [190]. It also inhibits influenza C and D viruses *in vitro* [191]. Baloxavir marboxil has been approved for clinical use [10, 192].

Hemagglutinin inhibitors. A broad-spectrum antiviral umifenovir interacts with influenza A and B hemagglutinin by inhibiting HA-mediated membrane fusion during influenza virus infection. Probably, umifenovir binds in a hydrophobic cavity in the HA trimer stem [193, 194]. A five-year double-blind, randomized, placebo-controlled phase IV clinical study of umifenovir was completed in 2016. The results confirmed the safety and effectiveness of umifenovir in influenza and other acute respiratory viral infections treatment in adults. The study was undertaken in Russia [195].

If the accessibility of endonuclease inhibitor baloxavir is increasing [175], other antivirals such as the polymerase inhibitor favipiravir [185, 186] and the HA inhibitor umifenovir [195] have limited availability [175].

Resistance to Direct-Acting Antivirals: Bad or Good?

A recent study [171] has made us wonder, do we always need to fight drug resistance? It was shown that serial passaging of mouse-adapted PR8 virus in the presence of camphor-based hemagglutinin inhibitor camphecene results in complete virus attenuation on the mouse model [171]. Six sequential passages of PR8 in MDCK cells in the presence of camphecene led to the detection of drug-resistant variants. However, unlike all other known drug-resistant viruses, camphecene-resistant strains completely lost the ability to cause lethal infection in mice. Serial passaging of PR8 in the presence of camphecene resulted in virus attenuation on the mouse model.

We can imagine a situation where widespread use of camphecene will lead to a rapidly increased circulation of camphecene-resistant influenza viruses. As a result of the replacement of camphecene-sensitive strains with camphecene-resistant ones, the virulence of circulating viruses for humans could be reduced.

Indirectly Acting Antivirals (Host-Directed Therapies)

The direct drugs are responsible for increasing the viral clearance rates being measured. Direct drugs are available on the market. In contrast, indirect drugs do not directly affect the virus life cycle; instead, such a drug affects a host cell (host-directed therapies). Most of the indirect drugs are under development.

As of today, a new indirect approach to the prevention and treatment of influenza infection is developing. This approach is based on the "virocell concept", which considers a virus-infected cell as a system with virus-specific metabolic changes following the plan for implementing the viral genome strategy [196 - 199]. According to this approach, the target of the pharmacological effect is not the virus but the metabolic systems of the host cell modified under the influence of a viral infection. Copper [200 - 202], lipid [203] or calcium metabolic system [204] could be used as a target.

The indirect impact on the virus through a change in its functional state may protect the host from influenza infection caused by any influenza virus strains, regardless of their sensitivity or resistance to widely used medications (such as rimantadine, oseltamivir, *etc*.). Further studies of drugs affecting infected cells' metabolism will help to define additional aspects of the pathogenesis of influenza infection and may offer additional directions for anti-influenza therapy development.

Some host-directed immunomodulators appear to cause the antiviral effect in treated individuals; for instance, macrolides may have immunomodulatory action, including downregulation of pro-inflammatory cytokines [205].

In contrast to the scheme presented in Fig. (**13**), Beigel and Hayden [175] distinguished four different kinds of recent clinical developments involving (**i**) selected polymerase inhibitors, (**ii**) neuraminidase inhibitors, (**iii**) antibody-based therapeutics, and (**iv**) host-directed therapies.

Antibody-Based Therapeutics

According to a study [175], antibody-based therapeutics including treatment with immune plasma, hyperimmune globulin, and anti-HA stem monoclonal antibodies can be considered as antiviral therapy.

CONCLUDING REMARKS

Since the 1918 pandemic, when the treatment of influenza was aimed at controlling only symptoms, a tremendous success in the prevention and treatment of this disease has been observed. A number of new vaccines and direct-acting influenza antivirals were licensed. However, the limited specificity of traditional vaccines and drug resistance of circulating viruses reduce the effectiveness of prophylaxis and treatment of influenza, requiring the development of new broad-spectrum preparations. Also, the treatment of severe influenza has not made such substantive advances, and according to Beigel *et al.* [175] "this is unlikely to change in the foreseeable future."

The high mutation rate of the influenza virus can dramatically alter virus antigenicity and reduce the effectiveness of vaccines. This problem can be avoided by developing long-lasting, broadly cross-reactive, and protective universal influenza virus vaccines. Besides, high-frequency viral mutation leads to the rapid development of drug resistance and reduces the effectiveness of antivirals. Here, combination therapy and indirectly acting drugs can be helpful.

REFERENCES

[1] WHO. Influenza (Seasonal): Fact Sheet N°211. November 2018. [Available from: http://www.who.int/mediacentre/factsheets/fs211/en/]

[2] Houser K, Subbarao K. Influenza vaccines: challenges and solutions. Cell Host Microbe 2015; 17(3): 295-300. [http://dx.doi.org/10.1016/j.chom.2015.02.012] [PMID: 25766291]

[3] Barberis I, Myles P, Ault SK, Bragazzi NL, Martini M. History and evolution of influenza control through vaccination: from the first monovalent vaccine to universal vaccines. J Prev Med Hyg 2016; 57(3): E115-20. [PMID: 27980374]

[4] Estrada LD, Schultz-Cherry S. Development of a universal influenza vaccine. J Immunol 2019; 202(2): 392-8. [http://dx.doi.org/10.4049/jimmunol.1801054] [PMID: 30617121]

[5] WHO. Global influenza preparedness plan. 2005. [Available from: http://www.who.int/csr/resources/publications/influenza/WHO_CDS_CSR_GIP_2005_5.pdf]

[6] WHO. Influenza updates. 2021. [Available from: https://www.who.int/influenza/surveillance_monitoring/updates/en/]

[7] ECDC. Scientific advice on seasonal influenza vaccination of children and pregnant women: Technical report. 2012. [Available from: https://www.ecdc.europa.eu/sites/portal/files/media/en/publications/Publications/Seasonal%20influenza%20vaccination%20of%20children%20and%20pregnant%20women.pdf]

[8] CDC. Influenza antiviral medications: Summary for clinicians. 2020. [Available from: https://www.cdc.gov/flu/professionals/antivirals/summary-clinicians.htm]

[9] CDC. Neuraminidase inhibitors for treatment of influenza A and B infections. 1999. [Available from: https://www.cdc.gov/mmwr/preview/mmwrhtml/rr4814a1.htm]

[10] Hayden FG, Shindo N. Influenza virus polymerase inhibitors in clinical development. Curr Opin Infect Dis 2019; 32(2): 176-86. [http://dx.doi.org/10.1097/QCO.0000000000000532] [PMID: 30724789]

[11] Hussain M, Galvin HD, Haw TY, Nutsford AN, Husain M. Drug resistance in influenza A virus: the epidemiology and management. Infect Drug Resist 2017; 10: 121-34. [http://dx.doi.org/10.2147/IDR.S105473] [PMID: 28458567]

[12] Smith W, Andrewes CH, Laidlaw PP. A virus obtained from influenza patients. Lancet 1933; 222(5732): 66-8. [http://dx.doi.org/10.1016/S0140-6736(00)78541-2]

[13] Smorodintseff AA, Tushinsky MD, Drobyshevskaja AI, Korovin AA, Osetroff AI. Investigation on volunteers infected with influenza virus. Am J Med Sci 1937; 194: 159-70. [http://dx.doi.org/10.1097/00000441-193708000-00002]

[14] Zhdanov VM, Aleksandrova GI, Gendon IuZ. Live influenza vaccine in the USSR: its development, study and practical use. Zh Mikrobiol Epidemiol Immunobiol 1986; 7(7): 3-13. [PMID: 3529762]

[15] Zhilova GP, Alexandrova GI, Zykov MP, Smorodintsev AA. Some problems of modern influenza prophylaxis with live vaccine. J Infect Dis 1977; 135(4): 681-6. [http://dx.doi.org/10.1093/infdis/135.4.681] [PMID: 853250]

[16] Francis T, Magill TP. The antibody response of human subjects vaccinated with the virus of human influenza. J Exp Med 1937; 65(2): 251-9.

[http://dx.doi.org/10.1084/jem.65.2.251] [PMID: 19870599]

[17] Goodpasture EW, Woodruff AM, Buddingh GJ. The cultivation of vaccine and other viruses in the chorioallantoic membrane of chick embryos. Science 1931; 74(1919): 371-2. [http://dx.doi.org/10.1126/science.74.1919.371] [PMID: 17810781]

[18] Montomoli E, Khadang B, Piccirella S, *et al.* Cell culture-derived influenza vaccines from Vero cells: a new horizon for vaccine production. Expert Rev Vaccines 2012; 11(5): 587-94. [http://dx.doi.org/10.1586/erv.12.24] [PMID: 22827244]

[19] Hardy CT, Young SA, Webster RG, Naeve CW, Owens RJ. Egg fluids and cells of the chorioallantoic membrane of embryonated chicken eggs can select different variants of influenza A (H3N2) viruses. Virology 1995; 211(1): 302-6. [http://dx.doi.org/10.1006/viro.1995.1405] [PMID: 7645225]

[20] Robertson JS, Cook P, Attwell AM, Williams SP. Replicative advantage in tissue culture of egg-adapted influenza virus over tissue-culture derived virus: implications for vaccine manufacture. Vaccine 1995; 13(16): 1583-8. [http://dx.doi.org/10.1016/0264-410X(95)00085-F] [PMID: 8578846]

[21] Williams SP, Robertson JS. Analysis of the restriction to the growth of nonegg-adapted human influenza virus in eggs. Virology 1993; 196(2): 660-5. [http://dx.doi.org/10.1006/viro.1993.1522] [PMID: 8372439]

[22] Robertson J. Clinical influenza virus and the embryonated hen's egg. Rev Med Virol 1993; 3: 97-106. [http://dx.doi.org/10.1002/rmv.1980030206]

[23] Ampofo WK, Baylor N, Cobey S, *et al.* Improving influenza vaccine virus selection: report of a WHO informal consultation held at WHO headquarters. Influenza Other Respir Virusess. Geneva, Switzerland 2010; 6: pp. (2)142-52. e1-5. [http://dx.doi.org/10.1038/nbt1261] [PMID: 17093488]

[24] Ulmer JB, Valley U, Rappuoli R. Vaccine manufacturing: challenges and solutions. Nat Biotechnol 2006; 24(11): 1377-83. [http://dx.doi.org/10.1038/nbt1261] [PMID: 17093488]

[25] Ampofo WK, Azziz-Baumgartner E, Bashir U, *et al.* Strengthening the influenza vaccine virus selection and development process: Report of the 3rd WHO Informal Consultation for Improving Influenza Vaccine Virus Selection held at WHO headquarters, Geneva, Switzerland, 1-3 April 2014. Vaccine 2015; 33(36): 4368-82. [http://dx.doi.org/10.1016/j.vaccine.2015.06.090] [PMID: 26148877]

[26] Pearce JM. Louis Pasteur and rabies: a brief note. J Neurol Neurosurg Psychiatry 2002; 73(1): 82. [http://dx.doi.org/10.1136/jnnp.73.1.82] [PMID: 12082056]

[27] Alexandrova GI, Smorodintsev AA. Obtaining of an additionally attenuated vaccinating cryophilic influenza strain. Rev Roum Inframicrobiol 1965; 2: 179-89.

[28] Maassab HF. Adaptation and growth characteristics of influenza virus at 25 degrees c. Nature 1967; 213(5076): 612-4. [http://dx.doi.org/10.1038/213612a0] [PMID: 6040602]

[29] Maassab HF, DeBorde DC. Development and characterization of cold-adapted viruses for use as live virus vaccines. Vaccine 1985; 3(5): 355-69. [http://dx.doi.org/10.1016/0264-410X(85)90124-0] [PMID: 3909681]

[30] Maassab HF. Plaque formation of influenza virus at 25 degrees C. Nature 1968; 219(5154): 645-6. [http://dx.doi.org/10.1038/219645a0] [PMID: 5665725]

[31] Burnet FM, Lind PE. A genetic approach to variation in influenza viruses; recombination of characters in influenza virus strains used in mixed infections. J Gen Microbiol 1951; 5(1): 59-66. [http://dx.doi.org/10.1099/00221287-5-1-59] [PMID: 14824471]

[32] Hirst GK, Gotlieb T. The experimental production of combination forms of virus. I. Occurrence of combination forms after simultaneous inoculation of the allantoic sac with two distinct strains of influenza virus. J Exp Med 1953; 98(1): 41-52.
[http://dx.doi.org/10.1084/jem.98.1.41] [PMID: 13069649]

[33] Kilbourne ED, Murphy JS. Genetic studies of influenza viruses. I. Viral morphology and growth capacity as exchangeable genetic traits. Rapid in ovo adaptation of early passage Asian strain isolates by combination with PR8. J Exp Med 1960; 111: 387-406.
[http://dx.doi.org/10.1084/jem.111.3.387] [PMID: 13755924]

[34] Wareing MD, Marsh GA, Tannock GA. Preparation and characterisation of attenuated cold-adapted influenza A reassortants derived from the A/Leningrad/134/17/57 donor strain. Vaccine 2002; 20(16): 2082-90.
[http://dx.doi.org/10.1016/S0264-410X(02)00056-7] [PMID: 11972977]

[35] Rudenko LG, Arden NH, Grigorieva E, *et al.* Immunogenicity and efficacy of Russian live attenuated and US inactivated influenza vaccines used alone and in combination in nursing home residents. Vaccine 2000; 19(2-3): 308-18.
[http://dx.doi.org/10.1016/S0264-410X(00)00153-5] [PMID: 10930686]

[36] Rudenko LG, Lonskaya NI, Klimov AI, Vasilieva RI, Ramirez A. Clinical and epidemiological evaluation of a live, cold-adapted influenza vaccine for 3-14-year-olds. Bull World Health Organ 1996; 74(1): 77-84.
[PMID: 8653819]

[37] Rudenko LG, Slepushkin AN, Monto AS, *et al.* Efficacy of live attenuated and inactivated influenza vaccines in schoolchildren and their unvaccinated contacts in Novgorod, Russia. J Infect Dis 1993; 168(4): 881-7.
[http://dx.doi.org/10.1093/infdis/168.4.881] [PMID: 8376833]

[38] Slepushkin AN, Obrosova-Serova NP, Burtseva EI, *et al.* Comparison of live attenuated and inactivated influenza vaccines in schoolchildren in Russia: I. Safety and efficacy in two Moscow schools, 1987/88. Vaccine 1993; 11(3): 323-8.
[http://dx.doi.org/10.1016/0264-410X(93)90194-3] [PMID: 8447161]

[39] Ambrose CS, Luke C, Coelingh K. Current status of live attenuated influenza vaccine in the United States for seasonal and pandemic influenza. Influenza Other Respir Viruses 2008; 2(6): 193-202.
[http://dx.doi.org/10.1111/j.1750-2659.2008.00056.x] [PMID: 19453395]

[40] Bergen R, Black S, Shinefield H, *et al.* Safety of cold-adapted live attenuated influenza vaccine in a large cohort of children and adolescents. Pediatr Infect Dis J 2004; 23(2): 138-44.
[http://dx.doi.org/10.1097/01.inf.0000109392.96411.4f] [PMID: 14872180]

[41] Neuzil KM, Zhu Y, Griffin MR, *et al.* Burden of interpandemic influenza in children younger than 5 years: a 25-year prospective study. J Infect Dis 2002; 185(2): 147-52.
[http://dx.doi.org/10.1086/338363] [PMID: 11807687]

[42] Zangwill KM, Belshe RB. Safety and efficacy of trivalent inactivated influenza vaccine in young children: a summary for the new era of routine vaccination. Pediatr Infect Dis J 2004; 23(3): 189-97.
[http://dx.doi.org/10.1097/01.inf.0000116292.46143.d6] [PMID: 15014289]

[43] Carter NJ, Curran MP. Live attenuated influenza vaccine (FluMist®; Fluenz™): a review of its use in the prevention of seasonal influenza in children and adults. Drugs 2011; 71(12): 1591-622.
[http://dx.doi.org/10.2165/11206860-000000000-00000] [PMID: 21861544]

[44] Mohn KG, Smith I, Sjursen H, Cox RJ. Immune responses after live attenuated influenza vaccination. Hum Vaccin Immunother 2018; 14(3): 571-8.
[http://dx.doi.org/10.1080/21645515.2017.1377376] [PMID: 28933664]

[45] WHO. Global influenza preparedness plan. 2005. [Available from: http://www.who.int/csr/resources/publications/influenza/WHO_CDS_CSR_GIP_2005_5.pdf]

[46] WHO. Global action plan to increase vaccine supply for influenza vaccines. 2006, Geneva, Belgium. 2006. [Available from: http://whqlibdoc.who.int/hq/2006/WHO_IVB_06.13_eng.pdf]

[47] Mallory RM, Malkin E, Ambrose CS, *et al.* Safety and immunogenicity following administration of a live, attenuated monovalent 2009 H1N1 influenza vaccine to children and adults in two randomized controlled trials. PLoS One 2010; 5(10): e13755. [http://dx.doi.org/10.1371/journal.pone.0013755] [PMID: 21060780]

[48] Rudenko L, van den Bosch H, Kiseleva I, *et al.* Live attenuated pandemic influenza vaccine: clinical studies on A/17/California/2009/38 (H1N1) and licensing of the Russian-developed technology to WHO for pandemic influenza preparedness in developing countries. Vaccine 2011; 29 (Suppl. 1): A40-4. [http://dx.doi.org/10.1016/j.vaccine.2011.04.122] [PMID: 21684428]

[49] WHO. Cumulative number of confirmed human cases for avian influenza A(H5N1) reported to World Health Organization, 2003-2020. [Available from: https://www.who.int/influenza/human_animal_interface/2020_01_20_tableH5N1.pdf?ua=1]

[50] Gao R, Cao B, Hu Y, *et al.* Human infection with a novel avian-origin influenza A (H7N9) virus. N Engl J Med 2013; 368(20): 1888-97. [http://dx.doi.org/10.1056/NEJMoa1304459] [PMID: 23577628]

[51] Chen Y, Liang W, Yang S, *et al.* Human infections with the emerging avian influenza A H7N9 virus from wet market poultry: clinical analysis and characterisation of viral genome. Lancet 2013; 381(9881): 1916-25. [http://dx.doi.org/10.1016/S0140-6736(13)60903-4] [PMID: 23623390]

[52] Neumann G, Noda T, Kawaoka Y. Emergence and pandemic potential of swine-origin H1N1 influenza virus. Nature 2009; 459(7249): 931-9. [http://dx.doi.org/10.1038/nature08157] [PMID: 19525932]

[53] Trucchi C, Paganino C, Amicizia D, *et al.* Universal influenza virus vaccines: what needs to happen next? Expert Opin Biol Ther 2019; 19(7): 671-83. [http://dx.doi.org/10.1080/14712598.2019.1604671] [PMID: 30957589]

[54] WHO. Availability of two new candidate reassortant vaccine viruses for pandemic (H1N1) 2009 virus vaccine development. 14 September 2009. [Available from: https://www.who.int/csr/resources/publications/swineflu/x_181_x_181a.pdf]

[55] Neurath AR, Rubin BA, Sillaman J, Tint H. The effect of nonaqueous solvents on the quaternary structure of viruses: a procedure for the simultaneous concentration, purification and disruption of influenza viruses. Microbios 1971; 4(14): 145-50. [PMID: 5162006]

[56] Laver WG, Webster RG. Preparation and immunogenicity of a purified influenza virus haemagglutinin and neuraminidase subunit vaccine. Postgrad Med J 1976; 52(608): 373-8. [http://dx.doi.org/10.1136/pgmj.52.608.373] [PMID: 785428]

[57] Beyer WE, Nauta JJ, Palache AM, Giezeman KM, Osterhaus AD. Immunogenicity and safety of inactivated influenza vaccines in primed populations: a systematic literature review and meta-analysis. Vaccine 2011; 29(34): 5785-92. [http://dx.doi.org/10.1016/j.vaccine.2011.05.040] [PMID: 21624411]

[58] Rajão DS, Pérez DR. Universal vaccines and vaccine platforms to protect against influenza viruses in humans and agriculture. Front Microbiol 2018; 9(123): 123. [http://dx.doi.org/10.3389/fmicb.2018.00123] [PMID: 29467737]

[59] Sedova E, Shcherbinin D, Migunov AI, *et al.* Recombinant influenza vaccines. Acta Naturae 2012; 4: 17-27. [http://dx.doi.org/10.32607/20758251-2012-4-4-17-27] [PMID: 23346377]

[60] Tosh PK, Poland GA. Emerging vaccines for influenza. Expert Opin Emerg Drugs 2008; 13(1): 21-40.

[http://dx.doi.org/10.1517/14728214.13.1.21] [PMID: 18321146]

[61] Wong SS, Webby RJ. Traditional and new influenza vaccines. Clin Microbiol Rev 2013; 26(3): 476-92.
[http://dx.doi.org/10.1128/CMR.00097-12] [PMID: 23824369]

[62] Fiore AE, Uyeki TM, Broder K, *et al.* Prevention and control of influenza with vaccines: recommendations of the Advisory Committee on Immunization Practices (ACIP). MMWR Recomm Rep 2010; 59(Rr-8): 1-62.

[63] Monto AS. Seasonal influenza vaccinations: specialized products for different target groups. Vaccine 2010; 28 (Suppl. 4): D14-23.
[http://dx.doi.org/10.1016/j.vaccine.2010.08.025] [PMID: 20713256]

[64] Krammer F. The human antibody response to influenza A virus infection and vaccination. Nat Rev Immunol 2019; 19(6): 383-97.
[http://dx.doi.org/10.1038/s41577-019-0143-6] [PMID: 30837674]

[65] Johnson PR, Feldman S, Thompson JM, Mahoney JD, Wright PF. Immunity to influenza A virus infection in young children: a comparison of natural infection, live cold-adapted vaccine, and inactivated vaccine. J Infect Dis 1986; 154(1): 121-7.
[http://dx.doi.org/10.1093/infdis/154.1.121] [PMID: 3711685]

[66] Wright PF, Thompson J, Vaughn WK, Folland DS, Sell SH, Karzon DT. Trials of influenza A/New Jersey/76 virus vaccine in normal children: an overview of age-related antigenicity and reactogenicity. J Infect Dis 1977; 136 (Suppl.): S731-41.
[http://dx.doi.org/10.1093/infdis/136.Supplement_3.S731] [PMID: 606798]

[67] Murphy BR, Rennels MB, Douglas RG, Jr, *et al.* Evaluation of influenza A/Hong Kong/123/77 (H1N1) ts-1A2 and cold-adapted recombinant viruses in seronegative adult volunteers. Infect Immun 1980; 29(2): 348-55.
[PMID: 7216417]

[68] Wright PF, Okabe N, McKee KT, Jr, Maassab HF, Karzon DT. Cold-adapted recombinant influenza A virus vaccines in seronegative young children. J Infect Dis 1982; 146(1): 71-9.
[http://dx.doi.org/10.1093/infdis/146.1.71] [PMID: 7045254]

[69] WHO. Recommendations to assure the quality, safety and efficacy of influenza vaccines (human, live attenuated) for intranasal administration. 2013. [Available from: http://www.who.int/biologicals/areas/vaccines/influenza/TRS_977_Annex_4.pdf?ua=1]

[70] Subbarao K, Barr I. A Tale of two mutations: Beginning to understand the problems with egg-based influenza vaccines? Cell Host Microbe 2019; 25(6): 773-5.
[http://dx.doi.org/10.1016/j.chom.2019.05.012] [PMID: 31194938]

[71] Wu NC, Lv H, Thompson AJ, *et al.* Preventing an antigenically disruptive mutation in egg-based H3N2 seasonal influenza vaccines by mutational incompatibility. Cell Host Microbe 2019; 25(6): 836-44.
[http://dx.doi.org/10.1016/j.chom.2019.04.013] [PMID: 31151913]

[72] Gubareva LV, Wood JM, Meyer WJ, *et al.* Codominant mixtures of viruses in reference strains of influenza virus due to host cell variation. Virology 1994; 199(1): 89-97.
[http://dx.doi.org/10.1006/viro.1994.1100] [PMID: 8116258]

[73] Kuwahara T, Takashita E, Fujisaki S, *et al.* Isolation of an egg-adapted influenza A(H3N2) virus without amino acid substitutions at the antigenic sites of its hemagglutinin. Jpn J Infect Dis 2018; 71(3): 234-8.
[http://dx.doi.org/10.7883/yoken.JJID.2017.551] [PMID: 29709975]

[74] Monto AS, Maassab HF, Bryan ER. Relative efficacy of embryonated eggs and cell culture for isolation of contemporary influenza viruses. J Clin Microbiol 1981; 13(1): 233-5.
[http://dx.doi.org/10.1128/JCM.13.1.233-235.1981] [PMID: 7462417]

[75] Gambaryan AS, Robertson JS, Matrosovich MN. Effects of egg-adaptation on the receptor-binding properties of human influenza A and B viruses. Virology 1999; 258(2): 232-9. [http://dx.doi.org/10.1006/viro.1999.9732] [PMID: 10366560]

[76] Gambaryan AS, Tuzikov AB, Piskarev VE, *et al.* Specification of receptor-binding phenotypes of influenza virus isolates from different hosts using synthetic sialylglycopolymers: non-egg-adapted human H1 and H3 influenza A and influenza B viruses share a common high binding affinity for 6′-sialyl(N-acetyllactosamine). Virology 1997; 232(2): 345-50. [http://dx.doi.org/10.1006/viro.1997.8572] [PMID: 9191848]

[77] Stevens J, Chen LM, Carney PJ, *et al.* Receptor specificity of influenza A H3N2 viruses isolated in mammalian cells and embryonated chicken eggs. J Virol 2010; 84(16): 8287-99. [http://dx.doi.org/10.1128/JVI.00058-10] [PMID: 20519409]

[78] WHO. Recommendations on the composition of influenza virus vaccines. 2020. [Available from: https://www.who.int/influenza/vaccines/virus/recommendations/en/]

[79] Milián E, Kamen AA. Current and emerging cell culture manufacturing technologies for influenza vaccines. BioMed Res Int 2015; 2015: 504831. [http://dx.doi.org/10.1155/2015/504831] [PMID: 25815321]

[80] Ng SK. Current cell-based influenza vaccine production technology as pandemic contingency. Hum Vaccin Immunother 2012; 8(2): 267-71. [http://dx.doi.org/10.4161/hv.18336] [PMID: 22426381]

[81] Youil R, Su Q, Toner TJ, *et al.* Comparative study of influenza virus replication in Vero and MDCK cell lines. J Virol Methods 2004; 120(1): 23-31. [http://dx.doi.org/10.1016/j.jviromet.2004.03.011] [PMID: 15234806]

[82] Nechaeva EA, Sen'kina TY, Ryzhikov AB, *et al.* Development of live cultural pandemic influenza vaccine Vector-Flu. BMC Proc 2011; 5 (Suppl. 8): 104. [http://dx.doi.org/10.1186/1753-6561-5-S8-P104] [PMID: 22373084]

[83] Kiseleva I, Su Q, Toner TJ, *et al.* Cell-based assay for the determination of temperature sensitive and cold adapted phenotypes of influenza viruses. J Virol Methods 2004; 116(1): 71-8. [http://dx.doi.org/10.1016/j.jviromet.2003.10.012] [PMID: 14715309]

[84] Moro PL, Winiecki S, Lewis P, Shimabukuro TT, Cano M. Surveillance of adverse events after the first trivalent inactivated influenza vaccine produced in mammalian cell culture (Flucelvax(®)) reported to the Vaccine Adverse Event Reporting System (VAERS), United States, 2013-2015. Vaccine 2015; 33(48): 6684-8. [http://dx.doi.org/10.1016/j.vaccine.2015.10.084] [PMID: 26518405]

[85] Nechaeva E, Ryzhikov A, Pyankova OG, *et al.* Study of immunogenicity and protective efficacy of live MDCK-derived pandemic influenza vaccine. Global J Infect Dis Clin Res 2019; 5: 010-5. [http://dx.doi.org/10.17352/2455-5363.000023] [PMID: 26518405]

[86] Palker T, Kiseleva I, Johnston K, *et al.* Protective efficacy of intranasal cold-adapted influenza A/New Caledonia/20/99 (H1N1) vaccines comprised of egg- or cell culture-derived reassortants. Virus Res 2004; 105(2): 183-94. [http://dx.doi.org/10.1016/j.virusres.2004.05.009] [PMID: 15351492]

[87] Morokutti A, Muster T, Ferko B. Intranasal vaccination with a replication-deficient influenza virus induces heterosubtypic neutralising mucosal IgA antibodies in humans. Vaccine 2014; 32(17): 1897-900. [http://dx.doi.org/10.1016/j.vaccine.2014.02.009] [PMID: 24560674]

[88] Romanova J, Krenn BM, Wolschek M, *et al.* Preclinical evaluation of a replication-deficient intranasal DeltaNS1 H5N1 influenza vaccine. PLoS One 2009; 4(6): e5984. [http://dx.doi.org/10.1371/journal.pone.0005984] [PMID: 19543385]

[89] Wacheck V, Egorov A, Groiss F, *et al.* A novel type of influenza vaccine: safety and immunogenicity

of replication-deficient influenza virus created by deletion of the interferon antagonist NS1. J Infect Dis 2010; 201(3): 354-62.
[http://dx.doi.org/10.1086/649428] [PMID: 20039806]

[90] Nachbagauer R, Palese P. Is a universal influenza virus vaccine possible? Annu Rev Med 2020; 71: 315-27.
[http://dx.doi.org/10.1146/annurev-med-120617-041310] [PMID: 31600454]

[91] Kim H, Webster RG, Webby RJ. Influenza virus: Dealing with a drifting and shifting pathogen. Viral Immunol 2018; 31(2): 174-83.
[http://dx.doi.org/10.1089/vim.2017.0141] [PMID: 29373086]

[92] Zost SJ, Parkhouse K, Gumina ME, *et al.* Contemporary H3N2 influenza viruses have a glycosylation site that alters binding of antibodies elicited by egg-adapted vaccine strains. Proc Natl Acad Sci USA 2017; 114(47): 12578-83.
[http://dx.doi.org/10.1073/pnas.1712377114] [PMID: 29109276]

[93] Kiseleva I. New points of departure for more global influenza vaccine use. Vaccines (Basel) 2020; 8(3): E410.
[http://dx.doi.org/10.3390/vaccines8030410] [PMID: 32718042]

[94] Fulvini AA, Ramanunninair M, Le J, *et al.* Gene constellation of influenza A virus reassortants with high growth phenotype prepared as seed candidates for vaccine production. PLoS One 2011; 6(6): e20823.
[http://dx.doi.org/10.1371/journal.pone.0020823] [PMID: 21695145]

[95] WHO. Availability of a new candidate reassortant vaccine virus for pandemic (H1N1) 2009 vaccine development. 18 March 2010. [Available from: https://www.who.int/csr/resources/publications /swineflu/candidate_niidrg_5.pdf]

[96] Kiseleva I, Larionova N, Rudenko L. Live attenuated reassortant vaccines based on A/Leningrad/134/17/57 master donor virus against H5 avian influenza. Open Microbiol J 2017; 11: 316-29.
[http://dx.doi.org/10.2174/1874285801711010316] [PMID: 29290844]

[97] Subbarao K, Webster RG, Kawaoka Y, Murphy BR. Are there alternative avian influenza viruses for generation of stable attenuated avian-human influenza A reassortant viruses? Virus Res 1995; 39(2-3): 105-18.
[http://dx.doi.org/10.1016/0168-1702(95)00082-8] [PMID: 8837878]

[98] Jazayeri SD, Poh CL. Development of universal influenza vaccines targeting conserved viral proteins. Vaccines (Basel) 2019; 7(4): E169.
[http://dx.doi.org/10.3390/vaccines7040169] [PMID: 31683888]

[99] Sautto GA, Kirchenbaum GA, Ross TM. Towards a universal influenza vaccine: different approaches for one goal. Virol J 2018; 15(1): 17.
[http://dx.doi.org/10.1186/s12985-017-0918-y] [PMID: 29370862]

[100] Xiang K, Ying G, Yan Z, *et al.* Progress on adenovirus-vectored universal influenza vaccines. Hum Vaccin Immunother 2015; 11(5): 1209-22.
[http://dx.doi.org/10.1080/21645515.2015.1016674] [PMID: 25876176]

[101] Soema PC, Kompier R, Amorij JP, Kersten GF. Current and next generation influenza vaccines: Formulation and production strategies. Eur J Pharm Biopharm 2015; 94: 251-63.
[http://dx.doi.org/10.1016/j.ejpb.2015.05.023] [PMID: 26047796]

[102] Zhang H, El Zowalaty ME. DNA-based influenza vaccines as immunoprophylactic agents toward universality. Future Microbiol 2016; 11(1): 153-64.
[http://dx.doi.org/10.2217/fmb.15.110] [PMID: 26673424]

[103] Zhang H, Wang L, Compans RW, Wang BZ. Universal influenza vaccines, a dream to be realized soon. Viruses 2014; 6(5): 1974-91.

[http://dx.doi.org/10.3390/v6051974] [PMID: 24784572]

[104] Zheng M, Luo J, Chen Z. Development of universal influenza vaccines based on influenza virus M and NP genes. Infection 2014; 42(2): 251-62. [http://dx.doi.org/10.1007/s15010-013-0546-4] [PMID: 24178189]

[105] Deng L, Cho KJ, Fiers W, Saelens X. M2e-based universal influenza A vaccines. Vaccines (Basel) 2015; 3(1): 105-36. [http://dx.doi.org/10.3390/vaccines3010105] [PMID: 26344949]

[106] Ebrahimi SM, Tebianian M. Influenza A viruses: why focusing on M2e-based universal vaccines. Virus Genes 2011; 42(1): 1-8. [http://dx.doi.org/10.1007/s11262-010-0547-7] [PMID: 21082230]

[107] Wei CJ, Crank MC, Shiver J, Graham BS, Mascola JR, Nabel GJ. Next-generation influenza vaccines: opportunities and challenges. Nat Rev Drug Discov 2020; 19(4): 239-52. [http://dx.doi.org/10.1038/s41573-019-0056-x] [PMID: 32060419]

[108] Fries LF, Smith GE, Glenn GM. A recombinant viruslike particle influenza A (H7N9) vaccine. N Engl J Med 2013; 369(26): 2564-6. [http://dx.doi.org/10.1056/NEJMc1313186] [PMID: 24224560]

[109] Kang SM, Kim MC, Compans RW. Virus-like particles as universal influenza vaccines. Expert Rev Vaccines 2012; 11(8): 995-1007. [http://dx.doi.org/10.1586/erv.12.70] [PMID: 23002980]

[110] Mezhenskaya D, Isakova-Sivak I, Rudenko L. M2e-based universal influenza vaccines: a historical overview and new approaches to development. J Biomed Sci 2019; 26(1): 76. [http://dx.doi.org/10.1186/s12929-019-0572-3] [PMID: 31629405]

[111] Wilson IA, Skehel JJ, Wiley DC. Structure of the haemagglutinin membrane glycoprotein of influenza virus at 3 A resolution. Nature 1981; 289(5796): 366-73. [http://dx.doi.org/10.1038/289366a0] [PMID: 7464906]

[112] Dormitzer PR. Rapid production of synthetic influenza vaccines. Curr Top Microbiol Immunol 2015; 386: 237-73. [http://dx.doi.org/10.1007/82_2014_399] [PMID: 24996863]

[113] Ulmer JB. Influenza DNA vaccines. Vaccine 2002; 20 (Suppl. 2): S74-6. [http://dx.doi.org/10.1016/S0264-410X(02)00136-6] [PMID: 12110264]

[114] Houser KV, Yamshchikov GV, Bellamy AR, *et al.* DNA vaccine priming for seasonal influenza vaccine in children and adolescents 6 to 17 years of age: A phase 1 randomized clinical trial. PLoS One 2018; 13(11): e0206837. [http://dx.doi.org/10.1371/journal.pone.0206837] [PMID: 30388160]

[115] Park KS, Seo YB, Lee JY, *et al.* Complete protection against a H5N2 avian influenza virus by a DNA vaccine expressing a fusion protein of H1N1 HA and M2e. Vaccine 2011; 29(33): 5481-7. [http://dx.doi.org/10.1016/j.vaccine.2011.05.062] [PMID: 21664216]

[116] Stachyra A, Góra-Sochacka A, Sirko A. DNA vaccines against influenza. Acta Biochim Pol 2014; 61(3): 515-22. [http://dx.doi.org/10.18388/abp.2014_1873] [PMID: 25210719]

[117] Hekele A, Bertholet S, Archer J, *et al.* Rapidly produced SAM(®) vaccine against H7N9 influenza is immunogenic in mice. Emerg Microbes Infect 2013; 2(8): e52. [http://dx.doi.org/10.1038/emi.2013.54] [PMID: 26038486]

[118] Brazzoli M, Magini D, Bonci A, *et al.* Induction of broad-based immunity and protective efficacy by self-amplifying mRNA vaccines encoding influenza virus hemagglutinin. J Virol 2015; 90(1): 332-44. [http://dx.doi.org/10.1128/JVI.01786-15] [PMID: 26468547]

[119] Magini D, Giovani C, Mangiavacchi S, *et al.* Self-amplifying mRNA vaccines expressing multiple

conserved influenza antigens confer protection against homologous and heterosubtypic viral challenge. PLoS One 2016; 11(8): e0161193.
[http://dx.doi.org/10.1371/journal.pone.0161193] [PMID: 27525409]

[120] Bahl K, Senn JJ, Yuzhakov O, *et al.* Preclinical and clinical demonstration of immunogenicity by mRNA vaccines against H10N8 and H7N9 influenza viruses 2017.
[http://dx.doi.org/10.1016/j.ymthe.2017.03.035]

[121] Assarsson E, Bui HH, Sidney J, *et al.* Immunomic analysis of the repertoire of T-cell specificities for influenza A virus in humans. J Virol 2008; 82(24): 12241-51.
[http://dx.doi.org/10.1128/JVI.01563-08] [PMID: 18842709]

[122] Quiñones-Parra S, Grant E, Loh L, *et al.* Preexisting CD8+ T-cell immunity to the H7N9 influenza A virus varies across ethnicities. Proc Natl Acad Sci USA 2014; 111(3): 1049-54.
[http://dx.doi.org/10.1073/pnas.1322229111] [PMID: 24395804]

[123] Wang TT, Tan GS, Hai R, *et al.* Vaccination with a synthetic peptide from the influenza virus hemagglutinin provides protection against distinct viral subtypes. Proc Natl Acad Sci USA 2010; 107(44): 18979-84.
[http://dx.doi.org/10.1073/pnas.1013387107] [PMID: 20956293]

[124] Staneková Z, Király J, Stropkovská A, *et al.* Heterosubtypic protective immunity against influenza A virus induced by fusion peptide of the hemagglutinin in comparison to ectodomain of M2 protein. Acta Virol 2011; 55(1): 61-7.
[http://dx.doi.org/10.4149/av_2011_01_61] [PMID: 21434706]

[125] Atsmon J, Kate-Ilovitz E, Shaikevich D, *et al.* Safety and immunogenicity of multimeric-001--a novel universal influenza vaccine. J Clin Immunol 2012; 32(3): 595-603.
[http://dx.doi.org/10.1007/s10875-011-9632-5] [PMID: 22318394]

[126] Atsmon J, Caraco Y, Ziv-Sefer S, *et al.* Priming by a novel universal influenza vaccine (Multimeric-001)-a gateway for improving immune response in the elderly population. Vaccine 2014; 32(44): 5816-23.
[http://dx.doi.org/10.1016/j.vaccine.2014.08.031] [PMID: 25173483]

[127] Krammer F, Palese P. Advances in the development of influenza virus vaccines. Nat Rev Drug Discov 2015; 14(3): 167-82.
[http://dx.doi.org/10.1038/nrd4529] [PMID: 25722244]

[128] Valero-Pacheco N, Pérez-Toledo M, Villasís-Keever MA, *et al.* Antibody persistence in adults two years after vaccination with an H1N1 2009 pandemic influenza virus-like particle vaccine. PLoS One 2016; 11(2): e0150146.
[http://dx.doi.org/10.1371/journal.pone.0150146] [PMID: 26919288]

[129] Ekiert DC, Bhabha G, Elsliger MA, *et al.* Antibody recognition of a highly conserved influenza virus epitope. Science 2009; 324(5924): 246-51.
[http://dx.doi.org/10.1126/science.1171491] [PMID: 19251591]

[130] Ni Y, Guo J, Turner D, Tizard I. An improved inactivated influenza vaccine with enhanced cross protection. Front Immunol 2018; 9: 1815.
[http://dx.doi.org/10.3389/fimmu.2018.01815] [PMID: 30140267]

[131] WHO. Global action plan to increase vaccine supply for influenza vaccines. 2006. [Available from: http://whqlibdoc.who.int/hq/2006/WHO_IVB_06.13_eng.pdf]

[132] Hoft DF, Babusis E, Worku S, *et al.* Live and inactivated influenza vaccines induce similar humoral responses, but only live vaccines induce diverse T-cell responses in young children. J Infect Dis 2011; 204(6): 845-53.
[http://dx.doi.org/10.1093/infdis/jir436] [PMID: 21846636]

[133] Mohn KG, Bredholt G, Brokstad KA, *et al.* Longevity of B-cell and T-cell responses after live attenuated influenza vaccination in children. J Infect Dis 2015; 211(10): 1541-9.

[http://dx.doi.org/10.1093/infdis/jiu654] [PMID: 25425696]

[134] Belshe RB, Gruber WC, Mendelman PM, *et al.* Efficacy of vaccination with live attenuated, cold-adapted, trivalent, intranasal influenza virus vaccine against a variant (A/Sydney) not contained in the vaccine. J Pediatr 2000; 136(2): 168-75.
[http://dx.doi.org/10.1016/S0022-3476(00)70097-7] [PMID: 10657821]

[135] Kiseleva I, Krutikova E, Stepanova E, *et al.* Cross-protective efficacy of monovalent live influenza B vaccines against genetically different lineages of B/Victoria and B/Yamagata in ferrets. BioMed Res Int 2018; 2018: 9695628.
[http://dx.doi.org/10.1155/2018/9695628] [PMID: 30246028]

[136] Rudenko L, Kiseleva I, Krutikova E, *et al.* Rationale for vaccination with trivalent or quadrivalent live attenuated influenza vaccines: Protective vaccine efficacy in the ferret model. PLoS One 2018; 13(12): e0208028.
[http://dx.doi.org/10.1371/journal.pone.0208028] [PMID: 30507951]

[137] Cheng X, Zengel JR, Suguitan AL Jr, *et al.* Evaluation of the humoral and cellular immune responses elicited by the live attenuated and inactivated influenza vaccines and their roles in heterologous protection in ferrets. J Infect Dis 2013; 208(4): 594-602.
[http://dx.doi.org/10.1093/infdis/jit207] [PMID: 23656978]

[138] Rekstin A, Isakova-Sivak I, Petukhova G, *et al.* Immunogenicity and cross protection in mice afforded by pandemic H1N1 live attenuated influenza vaccine containing wild-type nucleoprotein. BioMed Res Int 2017; 2017: 9359276.
[http://dx.doi.org/10.1155/2017/9359276] [PMID: 28210631]

[139] Jang YH, Seong BL. Cross-protective immune responses elicited by live attenuated influenza vaccines. Yonsei Med J 2013; 54(2): 271-82.
[http://dx.doi.org/10.3349/ymj.2013.54.2.271] [PMID: 23364956]

[140] Zens KD, Chen JK, Farber DL. Vaccine-generated lung tissue-resident memory T cells provide heterosubtypic protection to influenza infection. JCI Insight 2016; 1(10): 85832.
[http://dx.doi.org/10.1172/jci.insight.85832] [PMID: 27468427]

[141] Ambrose CS, Levin MJ, Belshe RB. The relative efficacy of trivalent live attenuated and inactivated influenza vaccines in children and adults. Influenza Other Respir Viruses 2011; 5(2): 67-75.
[http://dx.doi.org/10.1111/j.1750-2659.2010.00183.x] [PMID: 21306569]

[142] Hayward AC, Wang L, Goonetilleke N, *et al.* Natural T cell-mediated protection against seasonal and pandemic influenza. Results of the flu watch cohort study. Am J Respir Crit Care Med 2015; 191(12): 1422-31.
[http://dx.doi.org/10.1164/rccm.201411-1988OC] [PMID: 25844934]

[143] Bodewes R, Kreijtz JH, Rimmelzwaan GF. Yearly influenza vaccinations: a double-edged sword? Lancet Infect Dis 2009; 9(12): 784-8.
[http://dx.doi.org/10.1016/S1473-3099(09)70263-4] [PMID: 19879807]

[144] Song JM, Van Rooijen N, Bozja J, Compans RW, Kang SM. Vaccination inducing broad and improved cross protection against multiple subtypes of influenza A virus. Proc Natl Acad Sci USA 2011; 108(2): 757-61.
[http://dx.doi.org/10.1073/pnas.1012199108] [PMID: 21187388]

[145] Lee LY, Ha LA, Simmons C, *et al.* Memory T cells established by seasonal human influenza A infection cross-react with avian influenza A (H5N1) in healthy individuals. J Clin Invest 2008; 118(10): 3478-90.
[http://dx.doi.org/10.1172/JCI32460] [PMID: 18802496]

[146] Terajima M, Babon JAB, Co MDT, Ennis FA. Cross-reactive human B cell and T cell epitopes between influenza A and B viruses. Virol J 2013; 10(1): 244.
[http://dx.doi.org/10.1186/1743-422X-10-244] [PMID: 23886073]

[147] Toots M, Plemper RK. Next-generation direct-acting influenza therapeutics. Transl Res 2020; 220: 33-42.
[http://dx.doi.org/10.1016/j.trsl.2020.01.005] [PMID: 32088166]

[148] FDA. Influenza (Flu) antiviral drugs and related information. 2019. [Available from: https://www.fda.gov/drugs/information-drug-class/influenza-flu-antiviral-drus-and-related-information]

[149] Heo YA. Baloxavir: First global approval. Drugs 2018; 78(6): 693-7.
[http://dx.doi.org/10.1007/s40265-018-0899-1] [PMID: 29623652]

[150] McLaughlin MM, Skoglund EW, Ison MG. Peramivir: an intravenous neuraminidase inhibitor. Expert Opin Pharmacother 2015; 16(12): 1889-900.
[http://dx.doi.org/10.1517/14656566.2015.1066336] [PMID: 26153242]

[151] Scott LJ. Peramivir: A review in uncomplicated influenza. Drugs 2018; 78(13): 1363-70.
[http://dx.doi.org/10.1007/s40265-018-0981-8] [PMID: 30196350]

[152] Collier J. Zanamivir: an alternative translation. Philos Trans R Soc Lond B Biol Sci 2001; 356(1416): 1945-6.
[http://dx.doi.org/10.1098/rstb.2001.1039] [PMID: 11779395]

[153] Gubareva LV, Kaiser L, Hayden FG. Influenza virus neuraminidase inhibitors. Lancet 2000; 355(9206): 827-35.
[http://dx.doi.org/10.1016/S0140-6736(99)11433-8] [PMID: 10711940]

[154] Peramivir for influenza. Aust Prescr 2019; 42(4): 143.
[http://dx.doi.org/10.18773/austprescr.2019.047] [PMID: 31427848]

[155] Prichard WW, Hockessin D. Adamantanes and tricyclo[4.3.1.13.8] undecanes. US patent 3352912 1971. [Available from: https://www.freepatentsonline.com/3352912.pdf]

[156] Prichard WW, Hockessin D. Pharmaceutical compositions and methods of controlling influenza A virus infection utilizing substituted adamantanes and tricyclo[4.3.1.13.8] undecanes. US patent 3592934 1971. [Available from: https://www.freepatentsonline.com/3592934.pdf]

[157] Rabinovich S. Rimantadine therapy of influenza A infection in mice. Antimicrob Agents Chemother 1972; 1(5): 408-11.
[http://dx.doi.org/10.1128/AAC.1.5.408] [PMID: 4670482]

[158] Rabinovich S, Baldini JT, Bannister R. Treatment of influenza. The therapeutic efficacy of rimantadine HC1 in a naturally occurring influenza A2 outbreak. Am J Med Sci 1969; 257(5): 328-35.
[http://dx.doi.org/10.1097/00000441-196905000-00005] [PMID: 4891873]

[159] Polis J, Grava I. Method of producing alpha-methil-1-adamantylmethyl-amin hydrochloride. US Patent 3852352 1974. [Available from: https://patentimages.storage.googleapis.com/88/c2/dd/83195a073af5b7/US3852352.pdf]

[160] Pukite I. Talented but difficult. EY2010 Journalist Award, Latvia. 1974. [Available from: https://ec.europa.eu/employment_social/2010againstpoverty/export/sites/default/downloads/Journalist_Award/JA_LV.pdf]

[161] Dolin R, Reichman RC, Madore HP, Maynard R, Linton PN, Webber-Jones J. A controlled trial of amantadine and rimantadine in the prophylaxis of influenza A infection. N Engl J Med 1982; 307(10): 580-4.
[http://dx.doi.org/10.1056/NEJM198209023071002] [PMID: 7050702]

[162] Sears SD, Clements ML. Protective efficacy of low-dose amantadine in adults challenged with wild-type influenza A virus. Antimicrob Agents Chemother 1987; 31(10): 1470-3.
[http://dx.doi.org/10.1128/AAC.31.10.1470] [PMID: 3435099]

[163] Jing X, Ma C, Ohigashi Y, *et al.* Functional studies indicate amantadine binds to the pore of the influenza A virus M2 proton-selective ion channel. Proc Natl Acad Sci USA 2008; 105(31): 10967-72.

[http://dx.doi.org/10.1073/pnas.0804958105] [PMID: 18669647]

[164] Stouffer AL, Acharya R, Salom D, *et al.* Structural basis for the function and inhibition of an influenza virus proton channel. Nature 2008; 451(7178): 596-9. [http://dx.doi.org/10.1038/nature06528] [PMID: 18235504]

[165] Brown EG. Increased virulence of a mouse-adapted variant of influenza A/FM/1/47 virus is controlled by mutations in genome segments 4, 5, 7, and 8. J Virol 1990; 64(9): 4523-33. [http://dx.doi.org/10.1128/jvi.64.9.4523-4533.1990] [PMID: 2117072]

[166] Ilyushina NA, Khalenkov AM, Seiler JP, *et al.* Adaptation of pandemic H1N1 influenza viruses in mice. J Virol 2010; 84(17): 8607-16. [http://dx.doi.org/10.1128/JVI.00159-10] [PMID: 20592084]

[167] Magill TP, Francis T. The action of immune serum on human influenza virus *in vitro*. J Exp Med 1937; 65(6): 861-72. [http://dx.doi.org/10.1084/jem.65.6.861] [PMID: 19870640]

[168] Narasaraju T, Sim MK, Ng HH, *et al.* Adaptation of human influenza H3N2 virus in a mouse pneumonitis model: insights into viral virulence, tissue tropism and host pathogenesis. Microbes Infect 2009; 11(1): 2-11. [http://dx.doi.org/10.1016/j.micinf.2008.09.013] [PMID: 18983930]

[169] Prokopyeva EA, Sobolev IA, Prokopyev MV, Shestopalov AM. Adaptation of influenza A(H1N1)pdm09 virus in experimental mouse models. Infect Genet Evol 2016; 39: 265-71. [http://dx.doi.org/10.1016/j.meegid.2016.01.022] [PMID: 26829383]

[170] Taylor RM. Experimental infection with influenza A virus in mice: The increase in intrapulmonary virus after inoculation and the influence of various factors thereon. J Exp Med 1941; 73(1): 43-55. [http://dx.doi.org/10.1084/jem.73.1.43] [PMID: 19871064]

[171] Zarubaev VV, Pushkina EA, Borisevich SS, *et al.* Selection of influenza virus resistant to the novel camphor-based antiviral camphecene results in loss of pathogenicity. Virology 2018; 524: 69-77. [http://dx.doi.org/10.1016/j.virol.2018.08.011] [PMID: 30165308]

[172] Henle W, Henle G. Studies on the toxicity of influenza viruses. J Bacteriol 1946; 51: 632. [http://dx.doi.org/10.1084/jem.84.6.623] [PMID: 20987072]

[173] Kiseleva I, Rekstin A, Al Farroukh M, *et al.* Non-mouse-adapted H1N1pdm09 virus as a model for influenza research. Viruses 2020; 12(6): E590. [http://dx.doi.org/10.3390/v12060590] [PMID: 32485821]

[174] Cochran KW, Maassab HF, Tsunoda A, Berlin BS. Studies on the antiviral activity of amantadine hydrochloride. Ann N Y Acad Sci 1965; 130(1): 432-9. [http://dx.doi.org/10.1111/j.1749-6632.1965.tb12579.x] [PMID: 4285587]

[175] Beigel JH, Hayden FG. Influenza therapeutics in clinical practice-challenges and recent advances. Cold Spring Harb Perspect Med 2020. [http://dx.doi.org/10.1101/cshperspect.a038463] [PMID: 32041763]

[176] Uyeki TM, Bernstein HH, Bradley JS, *et al.* Clinical Practice Guidelines by the Infectious Diseases Society of America: 2018 Update on diagnosis, treatment, chemoprophylaxis, and institutional outbreak management of seasonal influenza. Clin Infect Dis 2019; 68(6): 895-902. [http://dx.doi.org/10.1093/cid/ciy874] [PMID: 30834445]

[177] Metlay JP, Waterer GW, Long AC, *et al.* Diagnosis and Treatment of Adults with Community-acquired Pneumonia. An Official Clinical Practice Guideline of the American Thoracic Society and Infectious Diseases Society of America. Am J Respir Crit Care Med 2019; 200(7): e45-67. [http://dx.doi.org/10.1164/rccm.201908-1581ST] [PMID: 31573350]

[178] Macdonald L. New influenza drugs zanamivir (Relenza) and oseltamivir (Tamiflu): unexpected serious reactions. CMAJ 2000; 163(7): 83-5. [PMID: 11033723]

[179] Kubo S, Tomozawa T, Kakuta M, Tokumitsu A, Yamashita M. Laninamivir prodrug CS-8958, a long-acting neuraminidase inhibitor, shows superior anti-influenza virus activity after a single administration. Antimicrob Agents Chemother 2010; 54(3): 1256-64. [http://dx.doi.org/10.1128/AAC.01311-09] [PMID: 20047917]

[180] Toyama K, Furuie H, Ishizuka H. Intrapulmonary pharmacokinetics of Laninamivir, a neuraminidase inhibitor, after a single nebulized administration of Laninamivir Octanoate in healthy japanese subjects. Antimicrob Agents Chemother 2017; 62(1): e01722-17. [http://dx.doi.org/10.1128/AAC.01722-17] [PMID: 29061751]

[181] Ishiguro N, Koseki N, Kaiho M, *et al.* Clinical effectiveness of four neuraminidase inhibitors (oseltamivir, zanamivir, laninamivir, and peramivir) for children with influenza A and B in the 2014-2015 to 2016-2017 influenza seasons in Japan. J Infect Chemother 2018; 24(6): 449-57. [http://dx.doi.org/10.1016/j.jiac.2018.01.013] [PMID: 29487035]

[182] Samson M, Pizzorno A, Abed Y, Boivin G. Influenza virus resistance to neuraminidase inhibitors. Antiviral Res 2013; 98(2): 174-85. [http://dx.doi.org/10.1016/j.antiviral.2013.03.014] [PMID: 23523943]

[183] Clark MP, Ledeboer MW, Davies I, *et al.* Discovery of a novel, first-in-class, orally bioavailable azaindole inhibitor (VX-787) of influenza PB2. J Med Chem 2014; 57(15): 6668-78. [http://dx.doi.org/10.1021/jm5007275] [PMID: 25019388]

[184] Hayden FG, Sugaya N, Hirotsu N, *et al.* Baloxavir marboxil for uncomplicated influenza in adults and adolescents. N Engl J Med 2018; 379(10): 913-23. [http://dx.doi.org/10.1056/NEJMoa1716197] [PMID: 30184455]

[185] Furuta Y, Komeno T, Nakamura T. Favipiravir (T-705), a broad spectrum inhibitor of viral RNA polymerase. Proc Jpn Acad, Ser B, Phys Biol Sci 2017; 93(7): 449-63. [http://dx.doi.org/10.2183/pjab.93.027] [PMID: 28769016]

[186] Delang L, Abdelnabi R, Neyts J. Favipiravir as a potential countermeasure against neglected and emerging RNA viruses. Antiviral Res 2018; 153: 85-94. [http://dx.doi.org/10.1016/j.antiviral.2018.03.003] [PMID: 29524445]

[187] Jin Z, Smith LK, Rajwanshi VK, Kim B, Deval J. The ambiguous base-pairing and high substrate efficiency of T-705 (Favipiravir) Ribofuranosyl 5′-triphosphate towards influenza A virus polymerase. PLoS One 2013; 8(7): e68347. [http://dx.doi.org/10.1371/journal.pone.0068347] [PMID: 23874596]

[188] Hayden F. Developing new antiviral agents for influenza treatment: what does the future hold? Clin Infect Dis 2009; 48 (Suppl. 1): S3-S13. [http://dx.doi.org/10.1086/591851] [PMID: 19067613]

[189] Chan-Tack KM, Murray JS, Birnkrant DB. Use of ribavirin to treat influenza. N Engl J Med 2009; 361(17): 1713-4. [http://dx.doi.org/10.1056/NEJMc0905290] [PMID: 19846864]

[190] Omoto S, Speranzini V, Hashimoto T, *et al.* Characterization of influenza virus variants induced by treatment with the endonuclease inhibitor baloxavir marboxil. Sci Rep 2018; 8(1): 9633. [http://dx.doi.org/10.1038/s41598-018-27890-4] [PMID: 29941893]

[191] Mishin VP, Patel MC, Chesnokov A, *et al.* Susceptibility of influenza A, B, C, and D viruses to baloxavir[1]. Emerg Infect Dis 2019; 25(10): 1969-72. [http://dx.doi.org/10.3201/eid2510.190607] [PMID: 31287050]

[192] Mifsud EJ, Hayden FG, Hurt AC. Antivirals targeting the polymerase complex of influenza viruses. Antiviral Res 2019; 169: 104545. [http://dx.doi.org/10.1016/j.antiviral.2019.104545] [PMID: 31247246]

[193] Leneva IA, Russell RJ, Boriskin YS, Hay AJ. Characteristics of arbidol-resistant mutants of influenza virus: implications for the mechanism of anti-influenza action of arbidol. Antiviral Res 2009; 81(2):

132-40.
[http://dx.doi.org/10.1016/j.antiviral.2008.10.009] [PMID: 19028526]

[194] Kadam RU, Wilson IA. Structural basis of influenza virus fusion inhibition by the antiviral drug Arbidol. Proc Natl Acad Sci USA 2017; 114(2): 206-14.
[http://dx.doi.org/10.1073/pnas.1617020114] [PMID: 28003465]

[195] Pshenichnaya NY, Bulgakova VA, Lvov NI, *et al.* Clinical efficacy of umifenovir in influenza and ARVI (study ARBITR). Ter Arkh 2019; 91(3): 56-63.
[http://dx.doi.org/10.26442/00403660.2019.03.000127] [PMID: 31094461]

[196] Rosenwasser S, Ziv C, Creveld SGV, Vardi A. Virocell metabolism: Metabolic innovations during host-virus interactions in the ocean. Trends Microbiol 2016; 24(10): 821-32.
[http://dx.doi.org/10.1016/j.tim.2016.06.006] [PMID: 27395772]

[197] Forterre P. The great virus comeback. Biol Aujourdhui 2013; 207(3): 153-68.
[http://dx.doi.org/10.1051/jbio/2013018] [PMID: 24330969]

[198] Forterre P. Viruses in the 21st century: From the curiosity-driven discovery of giant viruses to new concepts and definition of life. Clin Infect Dis 2017; 65(suppl_1): S74-9.
[http://dx.doi.org/10.1093/cid/cix349] [PMID: 28859344]

[199] Forterre P. To be or not to be alive: How recent discoveries challenge the traditional definitions of viruses and life. Stud Hist Philos Biol Biomed Sci 2016; 59: 100-8.
[http://dx.doi.org/10.1016/j.shpsc.2016.02.013] [PMID: 26996409]

[200] Puchkova LV, Babich PS, Zatulovskaia YA, Ilyechova EY, Di Sole F. Copper metabolism of newborns is adapted to milk ceruloplasmin as a nutritive source of copper: Overview of the current data. Nutrients 2018; 10(11): E1591.
[http://dx.doi.org/10.3390/nu10111591] [PMID: 30380720]

[201] Puchkova LV, Broggini M, Polishchuk EV, Ilyechova EY, Polishchuk RS. Silver ions as a tool for understanding different aspects of copper metabolism. Nutrients 2019; 11(6): E1364.
[http://dx.doi.org/10.3390/nu11061364] [PMID: 31213024]

[202] Rupp JC, Locatelli M, Grieser A, *et al.* Host Cell Copper Transporters CTR1 and ATP7A are important for Influenza A virus replication. Virol J 2017; 14(1): 11.
[http://dx.doi.org/10.1186/s12985-016-0671-7] [PMID: 28115001]

[203] Leser GP, Lamb RA. Influenza virus assembly and budding in raft-derived microdomains: a quantitative analysis of the surface distribution of HA, NA and M2 proteins. Virology 2005; 342(2): 215-27.
[http://dx.doi.org/10.1016/j.virol.2005.09.049] [PMID: 16249012]

[204] Zhou Y, Frey TK, Yang JJ. Viral calciomics: interplays between Ca^{2+} and virus. Cell Calcium 2009; 46(1): 1-17.
[http://dx.doi.org/10.1016/j.ceca.2009.05.005] [PMID: 19535138]

[205] Hung IFN, To KKW, Chan JFW, *et al.* Efficacy of clarithromycin-naproxen-oseltamivir combination in the treatment of patients hospitalized for influenza A(H3N2) infection: An open-label randomized, controlled, phase IIb/III trial. Chest 2017; 151(5): 1069-80.
[http://dx.doi.org/10.1016/j.chest.2016.11.012] [PMID: 27884765]

CHAPTER 5

Reasons for Insufficient Control of Influenza

Abstract: Unfortunately, influenza can not be controlled effectively. A number of factors are responsible for insufficient control of this disease. Viral ubiquity and transmissibility, a high spreading rate of the influenza virus, antigenic selection, antiviral treatment *etc.*, are responsible for the control of influenza. The main properties of the influenza virus described in the previous chapters that may influence the control of influenza are listed in this chapter.

Keywords: Control of influenza, Drug resistance, Genetic variability, Influenza virus, Transmission.

INTRODUCTION

Annually, influenza epidemics lead to significant morbidity, mortality, and social disruption. Seasonal influenza is responsible for 3-5 million severe illnesses worldwide and up to 650,000 respiratory deaths [1]. Influenza pandemics recur every 10–40 years. The number of deaths was estimated to be at least 50 million worldwide during the first virologically documented pandemic of Spanish flu, and 575,400 people died due to the last influenza pandemic H1N1pdm09 [2].

The claim of epidemic or pandemic potential of influenza virus rests on three premises. For a strain of influenza virus to cause the epidemic or pandemic, it should be (**i**) antigenically novel to the human immune system; (**ii**) virulent in the human host; (**iii**) person-to-person transmissible and ubiquitous (Fig. **1**) [3].

Over the past 35 years, a few number of the largest international forums for world-renowned experts in the field of influenza control, influenza vaccines, and related issues were dedicated to influenza held in different countries. Among them - ten Options for the control of influenza conferences, six The European Scientific Working group on Influenza (ESWI) conferences, seven Influenza Vaccines for the World (IVW) international conferences, *etc*. World-famous virologists from all over the world participated in these meetings to report their latest data and trends associated with current and new technologies and influenza vaccines.

Irina Kiseleva and Natalie Larionova

Unfortunately, despite all efforts and tremendous success in studying influenza, little progress in controlling this infection has been made so far. Why does the flu continue to be a poorly controlled infection? A number of factors are responsible for poor control of influenza and its treatment and cause of our failure to eradicate influenza worldwide. Eight main reasons that contributed to insufficient control for influenza are listed in Fig. (**2**).

WIDESPREAD OCCURRENCE OF INFLUENZA VIRUSES

Small microorganisms tend to have a cosmopolitan distribution [4]. Since the discovery of the first virus, the tobacco mosaic virus, in the 1890s, many viruses have been described and biologically characterized in the vast majority of regions of the planet, thus contributing to the concept of viral widespread occurrence. The abundance of viruses is coming to an astronomic number of viral particles on the planet.

Influenza viruses, as other respiratory viruses, are ubiquitous, and are found almost throughout the globe. Numerous studies evidenced that exposure to influenza virus is common and geographically widespread, potentially carrying a risk to susceptible populations [5].

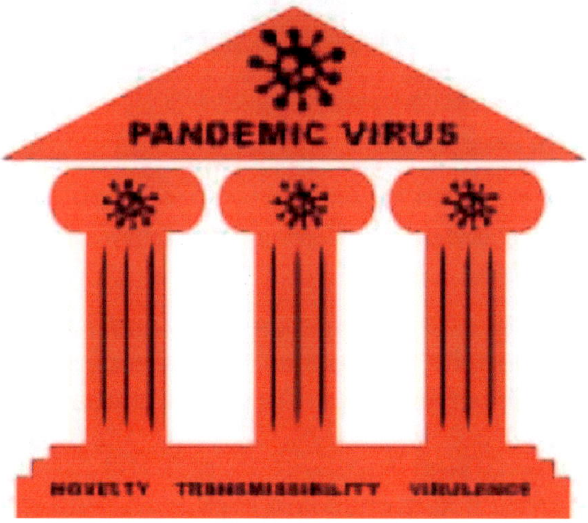

Fig. (1). Three pillars of an influenza outbreak. Novelty to the host immune system, transmissibility, and virulence in the host are required for the strain to be pandemic.

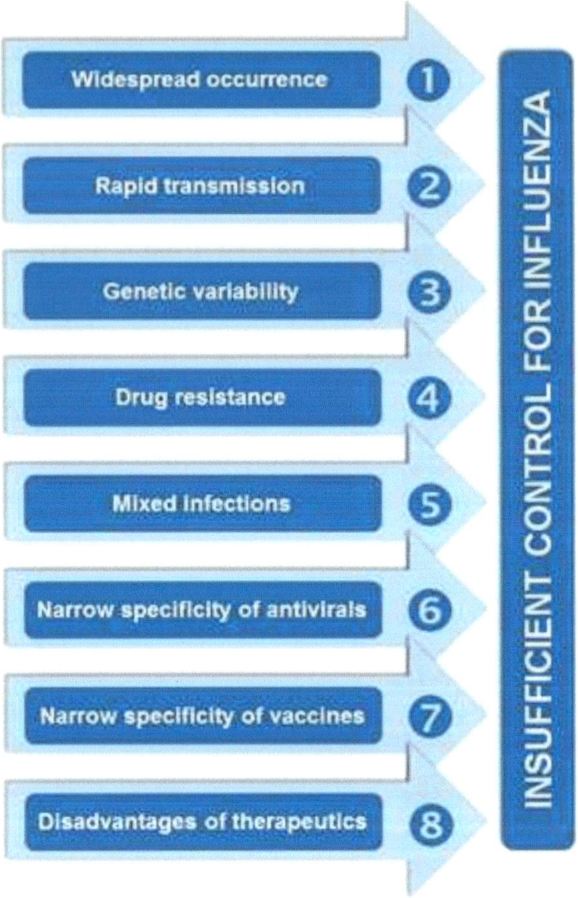

Fig. (2). Eight main reasons contributed to insufficient control for influenza.

The results of the ubiquity of the influenza virus and its high spreading rate are annual epidemics and sporadic pandemics. A full range of preventive measures should be taken to limit the effects of influenza infection and protect people (Fig. **3**).

The staggering widespread occurrence and high spreading rate are closely related to another important property of influenza virus – its transmissibility (contagiousness) which reinforcing the ubiquity of viruses on the planet.

Properties of influenza virus affecting its control	
Result of the specific property	**Necessary measures to be taken**
1. Widespread occurrence of influenza viruses	
Annual epidemics, sporadic pandemics as a result of high spreading rate of influenza virus	A full range of preventive measures
2. Contagiousness of influenza virus infection (rapid transmission)	
High spreading rate of influenza virus; high incidence	Special attention should be paid to risk groups (children, the elderly, medical workers, etc.)
3. High level of antigenic variability of influenza viruses	
Escaping from the immune response, reducing the effectiveness of vaccination	Using drugs with a wide range of therapeutic effects (e.g., interferons and their inducers), development of universal vaccines
4. Rapid development of drug resistance	
Decreased effectiveness of therapy	Monitoring for antiviral resistance; development of new drugs; combination chemotherapy
5. Mixed infections	
A single infection is caused by simultaneous viral-viral or viral-bacterial synergistic interaction	Using combination therapy
6. Narrow specificity of the action of most chemotherapy drugs	
Low efficiency of therapy for mixed infections	Using drugs with a wide range of therapeutic effects (e.g., interferons and their inducers)
7. Narrow specificity of the action of most conventional vaccines	
Low efficiency of vaccine therapy	Development of universal or broadly reactive vaccines
8. Disadvantages of current therapeutic methods	
Low efficiency of therapy; insufficient herd immunity	Achievement of at least 60 % herd immunity

Fig. (3). Attribution of causal relationships between properties of influenza virus and inadequate control for influenza.

RAPID TRANSMISSION

Influenza is one of the most contagious and rapidly spreading infectious diseases. Transmission (contagiousness) means capable of being able to be passed from one person to another. Influenza viruses spread from human to human *via* aerosols created by infected individuals. Infection occurs after an immunologically susceptible person inhales the aerosol contained influenza virus.

Influenza virus may be transmitted among humans in three different ways: (**i**) by direct contact with infected individuals; (**ii**) by indirect contact with surfaces in the immediate environment or with objects used by the infected person and contaminated with the virus; and (**iii**) by inhalation of virus-containing aerosols (Fig. **4**).

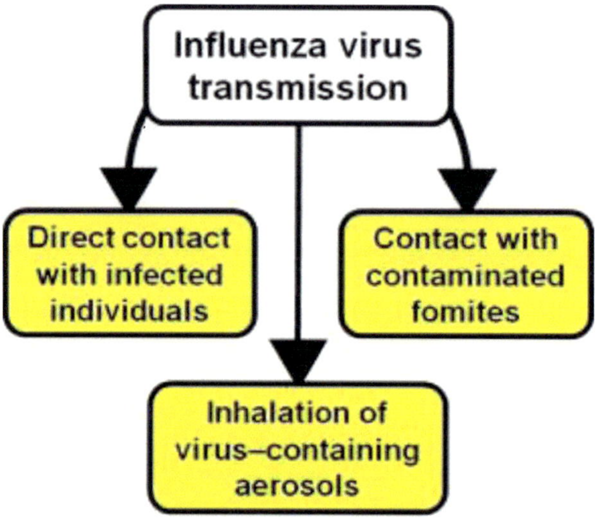

Fig. (4). Three main ways of influenza virus transmission.

Investigation of the transmission of human influenza viruses *via* environmental surfaces demonstrated that both influenza A and B viruses survived for 24-48 hrs on hard surfaces (stainless steel, plastic) and for less than 8-12 hrs. on cloth, paper, and tissues [6]. The authors suggested that virus transmission from an infected person could occur for 2-8 hrs. *via* hard surfaces and for a few minutes *via* paper tissues. Nevertheless, some believe that inhalation of virus-laden aerosols (airborne transmission) is generally thought to be the major way of influenza virus transmision. However, the transmission of influenza from human to human may result from the combination of these two ways.

In the context of animal influenza, the transmission of influenza viruses from poultry or pigs to humans appears to occur predominantly due to direct contact with infected animals [7, 8].

When an infected person sneezes, coughs, talks, sings, or whispers, they propel droplets of virus-contained mucus secretions into the air. The large droplets can land on surfaces within about one meter, while small droplets (aerosols) can be inhaled by susceptible people around. It has long been believed that the influenza viruses spread from human to human *via* aerosols created when an infected individual coughs, sneezes, *etc*. However, scientists from Maryland (the US) found that sneezing and coughing are not important for influenza virus aerosolization. They have shown that the infectious influenza virus can be transmitted from person to person by breathing, without coughing or sneezing [9]. When a person infected with a virus breathes, micron-sized aerosols containing millions of viral particles are released into the air. Interestingly, there are some amounts of influenza virus particles in the air constantly. Generally, these amounts are not sufficient to cause disease in humans. However, at a higher concentration of the airborne virus, the risk of human infection increases dramatically [10].

The result of rapid transmission of the influenza virus is high incidence. Special attention should be paid to risk groups (children, the elderly, medical workers, *etc.*) to protect them from a rapidly spreading infection.

GENETIC VARIABILITY

Despite years of surveillance and prevention measures, seasonal influenza viruses continue to cause epidemics yearly. The key process underlying these recurrent epidemics is the influenza virus rapid evolution. Punctuated antigenic changes, especially in HA, lead to escape from the immunity that was previously induced by infection or vaccination. Escape immunity [11] is a result of the high level of antigenic variability of viruses. A large body of literature describing influenza virus variability exists [12 - 17]. However, the timing and nature of the emergence of new strains remain unpredictable.

Antigenic Variability of Influenza Viruses

There are currently discovered 18 HA subtypes and 11 NA subtypes, most of which circulate in wild birds; but only three of them are known to have circulated widely in humans: H1N1, H2N2, and H3N2. H1N1 and H3N2 subtypes cause seasonal influenza virus epidemics. Circulating influenza B viruses belong to two major genetically distinct lineages (B/Victoria and B/Yamagata) (for more details, see Chapter **3**). Influenza viruses evolve through the gradual accumulation of mutations, which is a molecular basis of antigenic drift and genome reassortment (a phenomenon called antigenic shift).

New drifted antigenic variants of H3N2 viruses appear every 2–5 years. In contrast, new antigenic variants of H1N1 or influenza B viruses appear less frequently (3–8 years) [18 - 20]. Reassortment is also of evolutionary importance. Although most new genomic constellations will be unviable or deleterious, some of them may facilitate adaptation to new hosts, help evade host immune responses, and assist in the development of drug-resistance [11]. Reassortants of human, swine, or avian influenza viruses were detected [21, 22]. For instance, in 2010-2011, influenza H3N2 variant viruses (H3N2v) with the matrix (M) gene from the 2009 H1N1 pandemic virus were first identified in pigs and humans. Genetically distinct lineages of influenza virus B can co-circulate in the population and reassort [23, 24]. One of the most famous reassortants is the triple-reassortant that caused the 2009 pandemic [25].

Escaping from the Immune Response

Infection with an influenza virus induces innate and adaptive immune responses. The high variability of HA allows influenza viruses to escape from host immune surveillance and results in seasonal outbreaks. Viruses use various strategies to evade immune responses, resulting in the reduced clearing of the virus and virus-infected cells [26].

The evasion of influenza viruses from innate and adaptive immune responses reduces the effectiveness of vaccination. Escape immunity allows the virus to reinfect individuals who were once immune to the virus and necessitating reformulation of the seasonal influenza virus vaccine [27]. Vaccine formulations need to be updated every year to provide adequate protection. There is an urgent need for cross-protective or universal influenza vaccines to overcome the necessity for annual immunization against influenza. In recent years, a great amount of effort has been focused on developing a universal vaccine capable of eliciting broadly protective immunity (for more details, see Chapter **4**).

Our understanding of the complexity of interaction between host immune responses and influenza viruses has increased tremendously [26, 28 - 33]. However, there is still room for improvement in the development of broadly protective or universal vaccines.

The induction of cytotoxic T-lymphocyte responses to conserved epitopes of HA may be a venue to develop broadly protective vaccines [26]. Cross-neutralizing antibody responses directed to more conserved regions of the surface proteins are also attractive. Vaccines focused on the induction of antibodies against these regions would suffer to a lesser extent in terms of variability in the target region.

Ideally, the new vaccination strategies with broad protective and universal vaccines will counteract the variable nature of influenza viruses and protect against both seasonal influenza viruses and future pandemic strains.

DRUG RESISTANCE

At present, vaccination remains the major strategy for preventing seasonal and pandemic influenza [34]. Nevertheless, the role of effective antivirals in influenza treatment is substantial [35]. Influenza can be effectively treated with antivirals, especially in the pandemic situation when vaccine preparations are not immediately available. Current vaccine technology requires approximately 6 months lead-time [36]. This is a very long time for the preparation and release of a pandemic vaccine. The effectiveness of antivirals was demonstrated during the last rapid emergence of the pandemic H1N1pdm09 influenza virus when no vaccine was available.

For decades fusion inhibitors (M2 ion channel inhibitors) amantadine and rimantadine were the gold standard of therapy for the treatment and chemoprophylaxis of influenza A virus infection. Unfortunately, too many circulating strains of the influenza virus are resistant to these drugs. The emergence of pathogenic, transmissible, amantadine-resistant H1N1 [37], H1N1pdm09 [38], H3N2 [37] and some highly pathogenic H5N1 viruses [39] has limited the usefulness of M2 ion channel inhibitors [35, 37, 40 - 42]. Due to the high resistance (over 99%) CDC has no longer recommended amantadine and rimantadine for use since 2019-2020 influenza season [2, 43].

Currently, there are only two classes of antiviral drugs that are approved for use by FDA - neuraminidase blockers interfering with the release of progeny virus particles [43 - 46] and cap-dependent endonuclease inhibitors of the influenza virus polymerase PA subunit [46 - 48]. The development of drug-resistant strains and their global spread is of great clinical and public health concern and a major obstacle for the usefulness of these two classes of anti-influenza drugs [49]. The development of resistance of influenza viruses to antivirals occurs spontaneously or because of drug treatment of infected individuals. The high speed of spread of resistant strains suggests that they may have transmission advantages in untreated subjects.

Therefore, persistence monitoring for anti-influenza drug resistance is necessary. Limited efficiency of antivirals against recent resistant influenza virus strains and the increase in drug-resistance among the circulating strains of viruses have led to develop new therapeutic drugs and approaches. Of course, the development of new effective antivirals is the option, but it may take a few years between the first preliminary screening and the market.

An attractive approach for fighting drug resistance is so-called combination chemotherapy with concurrent use of two or more chemotherapy drugs with unrelated mechanisms of action and different mechanisms of drug resistance [35, 50]. These drugs may target (**i**) other viral proteins action (for instance, NA blocker in combination with ion channel blocker) or (**ii**) host factors for which the mechanisms of development of drug resistance differ (for instance, immunomodulator in combination with NA blocker).

The benefits of combination therapy is as follows: it may (**i**) reduce emergence of the resistance to a single drug and the effect of resistance that does emerge; (**ii**) enhance clinical outcomes and reduce the risk of respiratory complications, dose-related toxicity, side effects, and doses of individual drugs.

MIXED INFECTIONS

Different variations of interactions of circulating pathogens exist. It could be (**i**) single infection (one pathogen – one host) [51, 52]. However, not only mono infections could be found in nature. Mixed infections play a central role in the etiology of pulmonary diseases. The co-infection of more than one pathogen in one host is common across viruses and occurs more commonly than would be expected [53, 54]. The second variation of interactions between pathogens is (**ii**) mixed infection (co-infection), when two or more antigenically distinct pathogens infect one host [55 - 60]. The third variation is (**iii**) reassortment of two or more viruses in one host, when an individual is co-infected with two or more of the same type of virus, for example, influenza A viruses, resulting in reassortant progeny [61 - 63] (Fig. **5**).

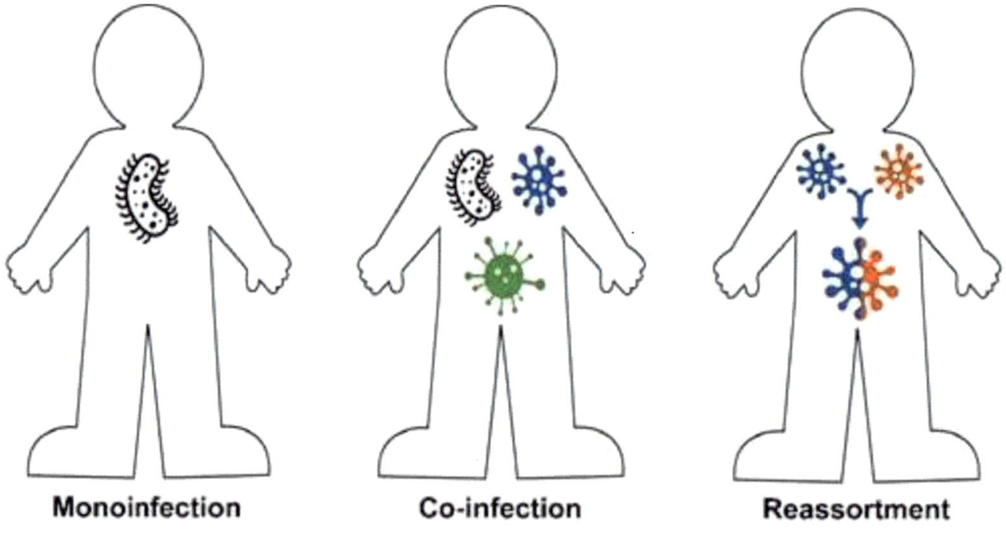

Fig. (5). Host-pathogen interactions [64]. The co-infection of one or more than one pathogen in one host is common across viruses.

Mixed infections significantly contribute to insufficient control for influenza. In the context of influenza, individuals may be co-infected with (**i**) another influenza virus of the same or different type/subtype; (**ii**) another ARD virus; (**iii**) bacterium (Fig. **6**). Each of these variants may negatively influence influenza treatment.

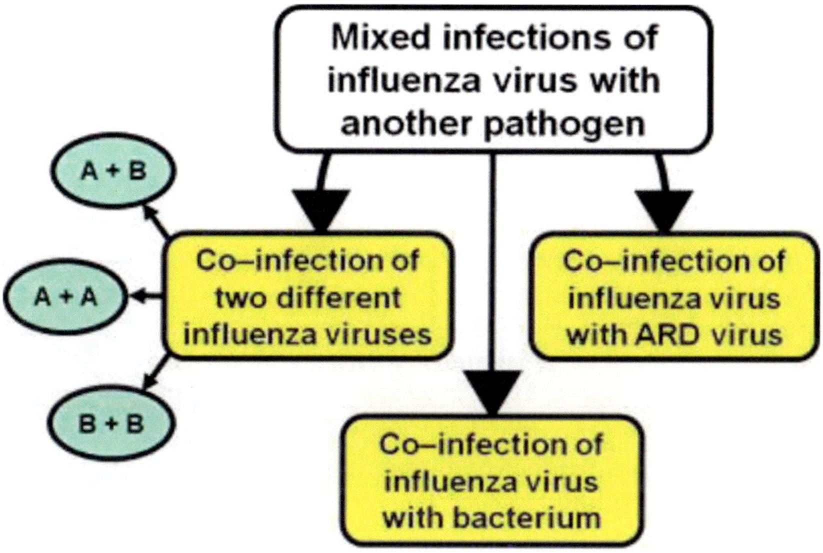

Fig. (6). The nature of mixed infections. Individuals infected with the influenza virus may be co-infected with another influenza or ARD virus or with a bacterium.

Viral-Viral Mixed Infections

The human respiratory tract hosts a diverse community of cocirculating viruses that are responsible for acute respiratory infections. In the respiratory airways, the ecological niche contained viruses and bacteria that periodically circulate and enter interspecific interactions that are formed [65 - 67]. The relationship between influenza virus and pneumococcus is an example of the interaction between viruses and bacteria, leading to about 100-fold increasing susceptibility to pneumococcal pneumonia after influenza infection [68]. This shared niche provides the opportunity for the virus–virus interactions [65]. During a seasonal rise in the incidence of influenza, a few antigenically distinct viruses circulate, the ratio of which varies annually. Seasonal influenza consists of variable mixes of influenza H3N2, H1N1pdm09, and the two B virus lineages.

Evidence of virus-virus interactions between 11 human respiratory viruses has recently been presented [65]. The experimental models of respiratory virus co-infections have demonstrated several interaction-induced effects, from enhanced or reduced viral growth to the attenuation of disease. In the case of recirculation, seasonal influenza-like virus and a common cold-like virus, a short-lived protective effect, such as that induced by interferon, was demonstrated [69, 70]. It has also been shown that cell fusion induced by certain viruses may enhance the replication of others in co-infections. However, despite epidemiological, clinical, and experimental indications of interactions among respiratory viruses, quantitatively robust evidence is still lacking [65].

The mixed infections of diverse influenza viruses are a common occurrence in humans [71]. Mixed infection of more than one influenza virus strain, types A+A [72 - 74] or types B+B [75], could lead to reassortment, which is an important mechanism for the evolution of influenza viruses which accelerates its rate (for more details see Chapter **3**).

Reassortment between seasonal H3N2 and H1N1pdm09 viruses was detected [74]. The H1N2 subtype viruses, which emerged during 2001, possessed an H1HA and seven genes related to H3N2 viruses [75]. Since the 1990s, widespread co-circulation of influenza B viruses of two genetic lineages has been the emergence of reassortants, represented by viruses, possessing a Victoria-lineage HA and a Yamagata-lineage NA or other way around [24, 75]. Reassortants may or may not acquire new properties that are not characteristic of parents. In the worst-case scenario, reassortment can generate a novel virus that causes disease and transmits between mammals [73].

Sequence analysis of isolates from individual influenza patients revealed a high level of a variety of mixed infection, including diverse lineages of the same influenza virus subtype, drug-resistant and drug-sensitive strains, those that are likely to differ in antigenicity, and even viruses of different influenza virus types (A and B). Thus, individuals can carry influenza viruses that differ in major phenotypic properties, including antigenic properties and sensitivity to antiviral agents [71].

When different respiratory viruses cocirculate, this can lead not only to cooperative but also to competitive forms of virus-virus interactions. It is believed that such interactions occur, for instance, among influenza and rhinoviruses [65]. It can also be assumed that competitive viruses are built between viruses, where an epidemically stronger rival wins. In such cases, the circulation of one viral pathogen may reduce the likelihood or delay infection by another. Thus, the pandemic H1N1pdm09 influenza virus delayed infection with RS-virus. The 2009 RS-virus infections were first noticed only when the infection with the pandemic swine influenza began to decline. Interestingly, in 2010–2011, infections with RS-virus were observed together with the pandemic swine influenza infection and may reflect the conversion of the pandemic influenza virus from a pandemic virus to a regular seasonal influenza virus [76]. In other cases, such interactions can have profound, aggravating the course of diseases, consequences.

In sum, the co-infection of more than one pathogen in one host is common across viruses and occurs more commonly than would be expected by chance in nature [53, 54].

Viral-Bacterial Mixed Infections

Mixed etiology of pulmonary infection is frequent, with pathogenic respiratory influenza-like viruses and bacteria co-infected in respiratory secretions. The clinical significance of these viral-bacterial co-infections has long been a controversial topic [77]. Viruses can alter the host's susceptibility to bacterial infections by altering both environmental conditions in the lung, which favor bacterial replication as well as by suppressing the host's defense mechanisms which prevent clearance of the bacteria [78]. Understanding the synergy or antagonism between viruses and bacteria will facilitate novel therapeutic approaches to prevent mortality associated with bacterial co-infections [79].

In healthy persons, seasonal influenza usually resolves with no consequences [80]. However, the risk for secondary bacterial pneumonia is high. Evidences for the role of the influenza virus in bacterial pneumonia are well-documented. Mixed infections of the airway are predicated on a bacterial superinfection either after, or concurrent with, an ongoing upper respiratory tract viral infection due to numerous viral ARD, including influenza A, influenza B [80].

Influenza is known as a risk factor for secondary bacterial pneumonia. Influenza viruses increase the susceptibility to secondary bacterial infections [81]. Severe bacterial pneumonia following influenza infection has been well described [67, 77, 80, 81]. Individuals suffering from viral pneumonia have at least a 40% chance of developing secondary post-influenza bacterial pneumonia [82]. Antibiotics are widely used to control secondary bacterial infections.

The expansion of drug-resistant viruses and antibiotic-resistant bacterial pathogens may lead to the inefficiency of modern treatment methods. The study of viral-bacterial interactions in pulmonary infection to develop better diagnostics, treatment, and prevention strategies is of particular importance.

A combination therapy that uses more than one medication may be a good option to treat mixed infections. Such polytherapy may be achieved by prescribing and administering separate drugs or, if available, dosage forms that contain more than one active ingredient.

NARROW SPECIFICITY OF ANTIVIRALS

Modern direct-acting antiviral compounds are of narrow specificity and inhibit the replication of influenza viruses only. However, there is another side to the coin. The narrow specificity of direct-acting antivirals has some disadvantages. Particularly, it leads to the low efficiency of therapy for mixed infections or infections caused by drug-resistant influenza virus strains.

An attractive approach to overcome the above-mentioned disadvantages of highly specific direct-acting antivirals is using combination therapy or indirectly acting drugs with a wide range of therapeutic effects (for instance, interferons and their inducers).

The value of the medical significance of interferons and their inducers is difficult to overestimate [83]. Interferons have broad-spectrum antiviral activities by blocking replication of viruses starting with influenza virus [84] and ending with SARS-CoV-2 [85]. Interferons are the first line of defense against both viruses and other pathogens. They act by inducing an antiviral response and mediating adaptive immune response; they are an important constituent element of innate immunity. Interferons are the essential part of a rational pharmacotherapy of a number of infectious and oncological diseases. For instance, interferons are widely used in the treatment of chronic hepatitis B and C, leukemia, Kaposi sarcoma, genital warts, lymphoma, melanoma, multiple sclerosis, *etc.*

Thus, by analogy with universal vaccines, ideally, universal broad-spectrum antivirals are needed.

NARROW SPECIFICITY OF CONVENTIONAL VACCINES

Due to the high level of antigenic variability of influenza viruses and their rapid mutation rate, vaccine composition has to be updated yearly to provide adequate protection [86]. The challenge of influenza vaccine development lies in the continually changing nature of the virus. Conventional influenza vaccines primarily induce virus-specific adaptive antibody responses and cannot protect against unexpected emerging influenza viruses [29, 86]. Inactivated conventional influenza vaccines are the most strain-specific and poorly effective against mismatched viruses [87, 88]. A mismatch between the vaccine virus and circulating strain leads to the low efficiency of vaccine therapy.

These problems can be solved by the development of broadly reactive vaccines and universal vaccines. Recently, great success has been made in the development of a universal vaccine capable of eliciting broadly protective immunity [89, 90].

Novel technologies that may improve influenza vaccines have been described. These technologies aim to develop universal vaccines eliciting long-lasting, broadly cross-reactive immune responses [89]. Major efforts to improve approaches for vaccine development are underway (for more details, see Chapter **4**).

Universal influenza vaccines seem to be the best option having the best potential to provide long-lasting protection against influenza viruses. However, the timeline for their development, as well as the true universality of protection they afford, remains uncertain [34]. While universal vaccines are still on the way, proponents of conventional vaccines are developing strategies to improve their efficiency because "A bird in the hand is better than two in the bush."

DISADVANTAGES OF CURRENT THERAPEUTIC AND DIAGNOSTIC METHODS

Disadvantages of current methods of influenza therapy associate with the majority of the above factors. A high level of antigenic variability of influenza viruses leads to a mismatch between circulating and vaccine viruses followed by escaping from the immune response; widespread use of direct-acting antivirals provokes the development of drug resistance. This, in turn, leads to low efficiency of therapy and insufficient herd immunity (Fig. **3**).

Herd Immunity

Herd immunity is a form of indirect protection from infection and occurs when a large percentage of a population has become immune to an infection, whether through previous infections or vaccination, thereby protecting those who are not immune [91 - 93].

60-90% of a community needs to become vaccinated against circulating influenza viruses to achieve herd immunity. The poor controllability of influenza is largely determined by the practical impossibility to reach a minimum required level of vaccine coverage of the population. Plans-Rubió [92] suggested that "the percentages of vaccination coverage registered in the United States and Europe are not sufficient to establish herd immunity."

For a more complete and accurate treatment, ideally, we need an individual approach - personalized medicine, currently used in such fields as oncology [94] or reproductive medicine [95]. Personalized medicine may prompt profound changes in the strategy of influenza therapy and the era of precision and personalized medicines will come.

Point-of-Care Tests

Besides, laboratory testing is required to distinguish infecting the influenza virus from other pathogens. The undiagnosed and untreated viral infection frequently becomes fatal for humans. Thus, early detection seems pivotal for accurate treatment, especially when vaccines and antivirals are not faultless [96]. Available rapid and accurate point-of-care tests that may allow for appropriate use of antiviral and antibiotic treatments and decrease patient illness are needed [97, 98].

The gold standard for rapid influenza diagnosis is molecular detection by RT-PCR assay, including the GeneXpert system, due to its sensitivity and specificity. Other routine assays used for influenza diagnosis include direct fluorescent immunoassay, enzyme immunoassay, viral culture, and some others [72, 96, 99] (for more details see Chapter **1**).

Ideally, an accurate point-of-care test should be rapid, easy to run and be performed directly "at the patient's bed" without transportation of specimens to the laboratory. Diagnosis using point-of-care tests will contribute to the appropriate prescription of treatments for influenza.

A standard complex of drugs is often prescribed. Empirical antiviral and antimicrobial use must be replaced with directed use of appropriate drugs, and influenza vaccine should be universal.

We are not pessimistic, but it looks like we are not able to be getting ahead of the influenza virus and predict its evolution, but what we can do is optimizing influenza therapy and prophylaxis. We sincerely believe that all the aforesaid is presenting new opportunities for better patient care.

CONCLUDING REMARKS

Taking into account all of the above information, soon, influenza is unlikely to be controllable. Overall, it is hard to control influenza. Some of the main underlying reasons for this are imperfect vaccine effectiveness, influenza cases do not always seek healthcare, and delayed and rare use of antiviral treatment. Nevertheless, the study on influenza and influenza viruses is still at its beginning. We cannot change the nature and properties of the influenza virus; therefore, we cannot control its ubiquity, transmissibility, genetic diversity and antigenic variability. However, what we can do is developing new effective therapeutic and prophylactic drugs.

REFERENCES

[1] WHO. Influenza (Seasonal): Fact Sheet N°211. November 2018. [Available from: http://www.who.int/mediacentre/factsheets/fs211/en/]

[2] CDC. Pandemic influenza. [Available from: https://www.cdc.gov/flu/pandemic-resources/index.htm]

[3] Lowen AC, Mubareka S, Tumpey TM, García-Sastre A, Palese P. The guinea pig as a transmission model for human influenza viruses. Proc Natl Acad Sci USA 2006; 103(26): 9988-92. [http://dx.doi.org/10.1073/pnas.0604157103] [PMID: 16785447]

[4] Fenchel T, Finlay BJ. The ubiquity of small species: Patterns of local and global diversity. Bioscience 2004; 54(8): 777. [http://dx.doi.org/10.1641/0006-3568(2004)054[0777:TUOSSP]2.0.CO;2]

[5] Kapoor S, Dhama K. Insight into influenza viruses of animals and humans. Cham Heidelberg New York Dordrecht London: Springer 2014.

[http://dx.doi.org/10.1007/978-3-319-05512-1]

[6] Bean B, Moore BM, Sterner B, Peterson LR, Gerding DN, Balfour HH, Jr. Survival of influenza viruses on environmental surfaces. J Infect Dis 1982; 146(1): 47-51. [http://dx.doi.org/10.1093/infdis/146.1.47] [PMID: 6282993]

[7] CDC. Information on avian influenza. 2020. [Available from: https://www.cdc.gov/flu/avianflu/index.htm]

[8] WHO. Influenza (Avian and other zoonotic). Key facts. 2018. [Available from: https://www.who.int/en/news-room/fact-sheets/detail/influenza-(avian-and-other-zoonotic]

[9] Yan J, Grantham M, Pantelic J, *et al.* Infectious virus in exhaled breath of symptomatic seasonal influenza cases from a college community. Proc Natl Acad Sci USA 2018; 115(5): 1081-6. [http://dx.doi.org/10.1073/pnas.1716561115] [PMID: 29348203]

[10] Nikitin N, Petrova E, Trifonova E, Karpova O. Influenza virus aerosols in the air and their infectiousness. Adv Virol 2014; 2014: 859090. [http://dx.doi.org/10.1155/2014/859090] [PMID: 25197278]

[11] Petrova VN, Russell CA. The evolution of seasonal influenza viruses. Nat Rev Microbiol 2018; 16(1): 47-60. [http://dx.doi.org/10.1038/nrmicro.2017.118] [PMID: 29081496]

[12] Nirmala J, Perez A, Culhane MR, Allerson MW, Sreevatsan S, Torremorell M. Genetic variability of influenza A virus in pigs at weaning in Midwestern United States swine farms. Transbound Emerg Dis 2021; 68(1): 62-75. [http://dx.doi.org/10.1111/tbed.13529] [PMID: 32187882]

[13] Duwell MM, Blythe D, Radebaugh MW, *et al.* Influenza A(H3N2) variant virus outbreak at three fairs - Maryland, 2017. MMWR Morb Mortal Wkly Rep 2018; 67(42): 1169-73. [http://dx.doi.org/10.15585/mmwr.mm6742a1] [PMID: 30359341]

[14] Yehia N, Hassan WMM, Sedeek A, Elhusseiny MH. Genetic variability of avian influenza virus subtype H5N8 in Egypt in 2017 and 2018. Arch Virol 2020; 165(6): 1357-66. [http://dx.doi.org/10.1007/s00705-020-04621-7] [PMID: 32285202]

[15] Wille M, Holmes EC. The ecology and evolution of influenza viruses. Cold Spring Harb Perspect Med 2019; 10(7): a038489. [http://dx.doi.org/10.1101/cshperspect.a038489] [PMID: 31871237]

[16] Xue KS, Bloom JD. Linking influenza virus evolution within and between human hosts. Virus Evol 2020; 6(1): veaa010. [http://dx.doi.org/10.1093/ve/veaa010] [PMID: 32082616]

[17] Rivas MJ, Alegretti M, Cóppola L, Ramas V, Chiparelli H, Goñi N. Epidemiology and genetic variability of circulating influenza B viruses in Uruguay, 2012-2019. Microorganisms 2020; 8(4): E591. [http://dx.doi.org/10.3390/microorganisms8040591] [PMID: 32325860]

[18] Chen H, Zhou X, Zheng J, Kwoh CK. Rules of co-occurring mutations characterize the antigenic evolution of human influenza A/H3N2, A/H1N1 and B viruses. BMC Med Genomics 2016; 9 (Suppl. 3): 69. [http://dx.doi.org/10.1186/s12920-016-0230-5] [PMID: 28117657]

[19] Chen R, Holmes EC. The evolutionary dynamics of human influenza B virus. J Mol Evol 2008; 66(6): 655-63. [http://dx.doi.org/10.1007/s00239-008-9119-z] [PMID: 18504518]

[20] Vijaykrishna D, Holmes EC, Joseph U, *et al.* The contrasting phylodynamics of human influenza B viruses. eLife 2015; 4: e05055. [http://dx.doi.org/10.7554/eLife.05055] [PMID: 25594904]

[21] Pinsent A, Fraser C, Ferguson NM, Riley S. A systematic review of reported reassortant viral lineages of influenza A. BMC Infect Dis 2016; 16: 3.
[http://dx.doi.org/10.1186/s12879-015-1298-9] [PMID: 26732146]

[22] Mehle A, Dugan VG, Taubenberger JK, Doudna JA. Reassortment and mutation of the avian influenza virus polymerase PA subunit overcome species barriers. J Virol 2012; 86(3): 1750-7.
[http://dx.doi.org/10.1128/JVI.06203-11] [PMID: 22090127]

[23] Tsai HP, Wang HC, Kiang D, *et al.* Increasing appearance of reassortant influenza B virus in Taiwan from 2002 to 2005. J Clin Microbiol 2006; 44(8): 2705-13.
[http://dx.doi.org/10.1128/JCM.02694-05] [PMID: 16891481]

[24] Tewawong N, Suntronwong N, Korkong S, Theamboonlers A, Vongpunsawad S, Poovorawan Y. Evidence for influenza B virus lineage shifts and reassortants circulating in Thailand in 2014-2016. Infect Genet Evol 2017; 47: 35-40.
[http://dx.doi.org/10.1016/j.meegid.2016.11.010] [PMID: 27845268]

[25] Vijaykrishna D, Poon LL, Zhu HC, *et al.* Reassortment of pandemic H1N1/2009 influenza A virus in swine. Science 2010; 328(5985): 1529.
[http://dx.doi.org/10.1126/science.1189132] [PMID: 20558710]

[26] van de Sandt CE, Kreijtz JH, Rimmelzwaan GF. Evasion of influenza A viruses from innate and adaptive immune responses. Viruses 2012; 4(9): 1438-76.
[http://dx.doi.org/10.3390/v4091438] [PMID: 23170167]

[27] Tricco AC, Chit A, Soobiah C, *et al.* Comparing influenza vaccine efficacy against mismatched and matched strains: a systematic review and meta-analysis. BMC Med 2013; 11: 153.
[http://dx.doi.org/10.1186/1741-7015-11-153] [PMID: 23800265]

[28] Chen X, Liu S, Goraya MU, Maarouf M, Huang S, Chen JL. Host immune response to influenza A virus infection. Front Immunol 2018; 9: 320.
[http://dx.doi.org/10.3389/fimmu.2018.00320] [PMID: 29556226]

[29] Krammer F. The human antibody response to influenza A virus infection and vaccination. Nat Rev Immunol 2019; 19(6): 383-97.
[http://dx.doi.org/10.1038/s41577-019-0143-6] [PMID: 30837674]

[30] Kreijtz JH, Fouchier RA, Rimmelzwaan GF. Immune responses to influenza virus infection. Virus Res 2011; 162(1-2): 19-30.
[http://dx.doi.org/10.1016/j.virusres.2011.09.022] [PMID: 21963677]

[31] Oltz EM. Immunity to influenza: Closing in on a moving target. J Immunol 2019; 202(2): 325-6.
[http://dx.doi.org/10.4049/jimmunol.1890024] [PMID: 30617112]

[32] Schmolke M, García-Sastre A. Evasion of innate and adaptive immune responses by influenza A virus. Cell Microbiol 2010; 12(7): 873-80.
[http://dx.doi.org/10.1111/j.1462-5822.2010.01475.x] [PMID: 20482552]

[33] Tate MD, Job ER, Deng YM, Gunalan V, Maurer-Stroh S, Reading PC. Playing hide and seek: how glycosylation of the influenza virus hemagglutinin can modulate the immune response to infection. Viruses 2014; 6(3): 1294-316.
[http://dx.doi.org/10.3390/v6031294] [PMID: 24638204]

[34] Harding AT, Heaton NS. Efforts to improve the seasonal influenza vaccine. Vaccines (Basel) 2018; 6(2): E19.
[http://dx.doi.org/10.3390/vaccines6020019] [PMID: 29601497]

[35] Govorkova EA, Webster RG. Combination chemotherapy for influenza. Viruses 2010; 2(8): 1510-29.
[http://dx.doi.org/10.3390/v2081510] [PMID: 21994692]

[36] Kandel R, Hartshorn KL. Novel strategies for prevention and treatment of influenza. Expert Opin Ther Targets 2005; 9(1): 1-22.

[http://dx.doi.org/10.1517/14728222.9.1.1] [PMID: 15757479]

[37] Deyde VM, Xu X, Bright RA, *et al.* Surveillance of resistance to adamantanes among influenza A(H3N2) and A(H1N1) viruses isolated worldwide. J Infect Dis 2007; 196(2): 249-57. [http://dx.doi.org/10.1086/518936] [PMID: 17570112]

[38] Cho HG, Choi JH, Kim WH, *et al.* High prevalence of amantadine-resistant influenza A virus isolated in Gyeonggi Province, South Korea, during 2005-2010. Arch Virol 2013; 158(1): 241-5. [http://dx.doi.org/10.1007/s00705-012-1482-9] [PMID: 23011311]

[39] Hurt AC, Selleck P, Komadina N, Shaw R, Brown L, Barr IG. Susceptibility of highly pathogenic A(H5N1) avian influenza viruses to the neuraminidase inhibitors and adamantanes. Antiviral Res 2007; 73(3): 228-31. [http://dx.doi.org/10.1016/j.antiviral.2006.10.004] [PMID: 17112602]

[40] Lampejo T. Influenza and antiviral resistance: An overview. Eur J Clin Microbiol Infect Dis 2020; 39(7): 1201-8. [http://dx.doi.org/10.1007/s10096-020-03840-9] [PMID: 32056049]

[41] Richman DD. Antiviral drug resistance. Antiviral Res 2006; 71(2-3): 117-21. [http://dx.doi.org/10.1016/j.antiviral.2006.03.004] [PMID: 16621040]

[42] Stilianakis NI, Perelson AS, Hayden FG. Drug resistance and influenza pandemics. Lancet 2002; 359(9320): 1862-3. [http://dx.doi.org/10.1016/S0140-6736(02)08691-9] [PMID: 12044407]

[43] CDC. Influenza antiviral medications: Summary for clinicians. 2020. [Available from: https://www.cdc.gov/flu/professionals/antivirals/summary-clinicians.htm]

[44] Collier J. Zanamivir: an alternative translation. Philos Trans R Soc Lond B Biol Sci 2001; 356(1416): 1945-6. [http://dx.doi.org/10.1098/rstb.2001.1039] [PMID: 11779395]

[45] Gubareva LV, Kaiser L, Hayden FG. Influenza virus neuraminidase inhibitors. Lancet 2000; 355(9206): 827-35. [http://dx.doi.org/10.1016/S0140-6736(99)11433-8] [PMID: 10711940]

[46] Peramivir for influenza. Aust Prescr 2019; 42(4): 143. [http://dx.doi.org/10.18773/austprescr.2019.047] [PMID: 31427848]

[47] Heo YA. Baloxavir: First global approval. Drugs 2018; 78(6): 693-7. [http://dx.doi.org/10.1007/s40265-018-0899-1] [PMID: 29623652]

[48] Hayden FG, Sugaya N, Hirotsu N, *et al.* Baloxavir marboxil for uncomplicated influenza in adults and adolescents. N Engl J Med 2018; 379(10): 913-23. [http://dx.doi.org/10.1056/NEJMoa1716197] [PMID: 30184455]

[49] Li TC, Chan MC, Lee N. Clinical implications of antiviral resistance in influenza. Viruses 2015; 7(9): 4929-44. [http://dx.doi.org/10.3390/v7092850] [PMID: 26389935]

[50] Yardley DA. Drug resistance and the role of combination chemotherapy in improving patient outcomes. Int J Breast Cancer 2013; 2013: 137414. [http://dx.doi.org/10.1155/2013/137414] [PMID: 23864953]

[51] Maurice JB, Patel A, Scott AJ, Patel K, Thursz M, Lemoine M. Prevalence and risk factors of nonalcoholic fatty liver disease in HIV-monoinfection. AIDS 2017; 31(11): 1621-32. [http://dx.doi.org/10.1097/QAD.0000000000001504] [PMID: 28398960]

[52] Navarro-Marí JM, Pérez-Ruiz M, Galán Montemayor JC, *et al.* Circulation of other respiratory viruses and viral co-infection during the 2009 pandemic influenza. Enferm Infec Microbiol Clin 2012; 30 (Suppl. 4): 25-31. [http://dx.doi.org/10.1016/S0213-005X(12)70101-5] [PMID: 23116789]

[53] Díaz-Muñoz SL. Viral coinfection is shaped by host ecology and virus-virus interactions across diverse microbial taxa and environments. Virus Evol 2017; 3(1): vex011. [http://dx.doi.org/10.1093/ve/vex011] [PMID: 28469939]

[54] Alcaide C, Rabadán MP, Moreno-Pérez MG, Gómez P. Implications of mixed viral infections on plant disease ecology and evolution. Adv Virus Res 2020; 106: 145-69. [http://dx.doi.org/10.1016/bs.aivir.2020.02.001] [PMID: 32327147]

[55] Nowak MD, Sordillo EM, Gitman MR, Paniz Mondolfi AE. Coinfection in SARS-CoV-2 infected patients: Where are influenza virus and rhinovirus/enterovirus? J Med Virol 2020; 92(10): 1699-700. [http://dx.doi.org/10.1002/jmv.25953] [PMID: 32352574]

[56] Ou X, Zhou L, Huang H, Lin Y, Pan X, Chen D. A severe case with co-infection of SARS-CoV-2 and common respiratory pathogens. Travel Med Infect Dis 2020; 35: 101672. [http://dx.doi.org/10.1016/j.tmaid.2020.101672] [PMID: 32305630]

[57] Ozaras R, Cirpin R, Duran A, *et al.* Influenza and COVID-19 co-infection: Report of 6 cases and review of the Literature. J Med Virol 2020; 92(11): 2657-65. [http://dx.doi.org/10.1002/jmv.26125] [PMID: 32497283]

[58] Rawson TM, Moore LSP, Zhu N, *et al.* Bacterial and fungal co-infection in individuals with coronavirus: A rapid review to support COVID-19 antimicrobial prescribing. Clin Infect Dis 2020; 71(9): 2459-68. [http://dx.doi.org/10.1093/cid/ciaa530] [PMID: 32358954]

[59] Rodriguez JA, Rubio-Gomez H, Roa AA, Miller N, Eckardt PA. Co-Infection with SARS-COV-2 and Parainfluenza in a young adult patient with pneumonia: Case report. IDCases 2020; 20: e00762. [http://dx.doi.org/10.1016/j.idcr.2020.e00762] [PMID: 32368493]

[60] Touzard-Romo F, Tapé C, Lonks JR. Co-infection with SARS-CoV-2 and human metapneumovirus. Rhode Isl Med J (2013) 2020; 103(2): 75-6. [PMID: 32192233]

[61] Steel J, Lowen AC. Influenza A virus reassortment. Curr Top Microbiol Immunol 2014; 385: 377-401. [http://dx.doi.org/10.1007/82_2014_395] [PMID: 25007845]

[62] Klempa B. Reassortment events in the evolution of hantaviruses. Virus Genes 2018; 54(5): 638-46. [http://dx.doi.org/10.1007/s11262-018-1590-z] [PMID: 30047031]

[63] Matsuzaki Y, Sugawara K, Takashita E, *et al.* Genetic diversity of influenza B virus: the frequent reassortment and cocirculation of the genetically distinct reassortant viruses in a community. J Med Virol 2004; 74(1): 132-40. [http://dx.doi.org/10.1002/jmv.20156] [PMID: 15258979]

[64] Kiseleva I, Grigorieva E, Larionova N, Al Farroukh M, Rudenko L. COVID-19 in light of seasonal respiratory infections. Biology (Basel) 2020; 9(9): 240. [http://dx.doi.org/10.3390/biology9090240] [PMID: 32825427]

[65] Nickbakhsh S, Mair C, Matthews L, *et al.* Virus-virus interactions impact the population dynamics of influenza and the common cold. Proc Natl Acad Sci USA 2019; 116(52): 27142-50. [http://dx.doi.org/10.1073/pnas.1911083116] [PMID: 31843887]

[66] Lloyd-Smith JO. Vacated niches, competitive release and the community ecology of pathogen eradication. Philos Trans R Soc Lond B Biol Sci 2013; 368(1623): 20120150. [http://dx.doi.org/10.1098/rstb.2012.0150] [PMID: 23798698]

[67] Bosch AA, Biesbroek G, Trzcinski K, Sanders EA, Bogaert D. Viral and bacterial interactions in the upper respiratory tract. PLoS Pathog 2013; 9(1): e1003057. [http://dx.doi.org/10.1371/journal.ppat.1003057] [PMID: 23326226]

[68] Shrestha S, Foxman B, Weinberger DM, Steiner C, Viboud C, Rohani P. Identifying the interaction between influenza and pneumococcal pneumonia using incidence data. Sci Transl Med 2013; 5(191):

191ra84.
[http://dx.doi.org/10.1126/scitranslmed.3005982] [PMID: 23803706]

[69] Isaacs A, Burke DC. Viral interference and interferon. Br Med Bull 1959; 15: 185-8. [http://dx.doi.org/10.1093/oxfordjournals.bmb.a069760] [PMID: 13853040]

[70] Dianzani F. Viral interference and interferon. Ric Clin Lab 1975; 5(3): 196-213. [PMID: 778995]

[71] Ghedin E, Fitch A, Boyne A, *et al.* Mixed infection and the genesis of influenza virus diversity. J Virol 2009; 83(17): 8832-41. [http://dx.doi.org/10.1128/JVI.00773-09] [PMID: 19553313]

[72] Tao H, Li L, White MC, Steel J, Lowen AC. Influenza A virus coinfection through transmission can support high levels of reassortment. J Virol 2015; 89(16): 8453-61. [http://dx.doi.org/10.1128/JVI.01162-15] [PMID: 26041285]

[73] Williams GD, Pinto AK, Doll B, Boon ACM. A North American H7N3 influenza virus supports reassortment with 2009 pandemic H1N1 and induces disease in mice without prior adaptation. J Virol 2016; 90(9): 4796-806. [http://dx.doi.org/10.1128/JVI.02761-15] [PMID: 26937034]

[74] Phipps KL, Marshall N, Tao H, *et al.* Seasonal H3N2 and 2009 pandemic H1N1 influenza A viruses reassort efficiently but produce attenuated progeny. J Virol 2017; 91(17): e00830-17. [http://dx.doi.org/10.1128/JVI.00830-17] [PMID: 28637755]

[75] Lin YP, Gregory V, Bennett M, Hay A. Recent changes among human influenza viruses. Virus Res 2004; 103(1-2): 47-52. [http://dx.doi.org/10.1016/j.virusres.2004.02.011] [PMID: 15163487]

[76] Hirsh S, Hindiyeh M, Kolet L, *et al.* Epidemiological changes of respiratory syncytial virus (RSV) infections in Israel. PLoS One 2014; 9(3): e90515. [http://dx.doi.org/10.1371/journal.pone.0090515] [PMID: 24594694]

[77] Brealey JC, Sly PD, Young PR, Chappell KJ. Viral bacterial co-infection of the respiratory tract during early childhood. FEMS Microbiol Lett 2015; 362(10): fnv062. [http://dx.doi.org/10.1093/femsle/fnv062] [PMID: 25877546]

[78] Babiuk LA, Lawman MJ, Ohmann HB. Viral-bacterial synergistic interaction in respiratory disease. Adv Virus Res 1988; 35: 219-49. [http://dx.doi.org/10.1016/S0065-3527(08)60713-7] [PMID: 3148270]

[79] Jia L, Xie J, Zhao J, *et al.* Mechanisms of severe mortality-associated bacterial co-infections following influenza virus infection. Front Cell Infect Microbiol 2017; 7: 338. [http://dx.doi.org/10.3389/fcimb.2017.00338] [PMID: 28824877]

[80] Bakaletz LO. Viral-bacterial co-infections in the respiratory tract. Curr Opin Microbiol 2017; 35: 30-5. [http://dx.doi.org/10.1016/j.mib.2016.11.003] [PMID: 27940028]

[81] Bellinghausen C, Rohde GGU, Savelkoul PHM, Wouters EFM, Stassen FRM. Viral-bacterial interactions in the respiratory tract. J Gen Virol 2016; 97(12): 3089-102. [http://dx.doi.org/10.1099/jgv.0.000627] [PMID: 27902340]

[82] Jakab GJ. Viral-bacterial interactions in pulmonary infection. Adv Vet Sci Comp Med 1982; 26: 155-71. [PMID: 6758537]

[83] Ershov FI. The medical significance of interferons and their inducers. Vestn Ross Akad Med Nauk 2004; (2): 9-13. [PMID: 15101201]

[84] Mannering GJ, Deloria LB. The pharmacology and toxicology of the interferons: an overview. Annu Rev Pharmacol Toxicol 1986; 26: 455-515.

[http://dx.doi.org/10.1146/annurev.pa.26.040186.002323] [PMID: 2424363]

[85] Mantlo E, Bukreyeva N, Maruyama J, Paessler S, Huang C. Antiviral activities of type I interferons to SARS-CoV-2 infection. Antiviral Res 2020; 179: 104811. [http://dx.doi.org/10.1016/j.antiviral.2020.104811] [PMID: 32360182]

[86] Kim H, Webster RG, Webby RJ. Influenza virus: Dealing with a drifting and shifting pathogen. Viral Immunol 2018; 31(2): 174-83. [http://dx.doi.org/10.1089/vim.2017.0141] [PMID: 29373086]

[87] Ekiert DC, Bhabha G, Elsliger MA, *et al.* Antibody recognition of a highly conserved influenza virus epitope. Science 2009; 324(5924): 246-51. [http://dx.doi.org/10.1126/science.1171491] [PMID: 19251591]

[88] Rajão DS, Pérez DR. Universal vaccines and vaccine platforms to protect against influenza viruses in humans and agriculture. Front Microbiol 2018; 9(123): 123. [http://dx.doi.org/10.3389/fmicb.2018.00123] [PMID: 29467737]

[89] Nachbagauer R, Palese P. Is a universal influenza virus vaccine possible? Annu Rev Med 2020; 71: 315-27. [http://dx.doi.org/10.1146/annurev-med-120617-041310] [PMID: 31600454]

[90] Ostrowsky J, Arpey M, Moore K, *et al.* Tracking progress in universal influenza vaccine development. Curr Opin Virol 2020; 40: 28-36. [http://dx.doi.org/10.1016/j.coviro.2020.02.003] [PMID: 32279026]

[91] Metcalf CJE, Ferrari M, Graham AL, Grenfell BT. Understanding herd immunity. Trends Immunol 2015; 36(12): 753-5. [http://dx.doi.org/10.1016/j.it.2015.10.004] [PMID: 26683689]

[92] Plans-Rubió P. The vaccination coverage required to establish herd immunity against influenza viruses. Prev Med 2012; 55(1): 72-7. [http://dx.doi.org/10.1016/j.ypmed.2012.02.015] [PMID: 22414740]

[93] Fine P, Eames K, Heymann DL. "Herd immunity": a rough guide. Clin Infect Dis 2011; 52(7): 911-6. [http://dx.doi.org/10.1093/cid/cir007] [PMID: 21427399]

[94] Baker SG, Sargent DJ. Designing a randomized clinical trial to evaluate personalized medicine: a new approach based on risk prediction. J Natl Cancer Inst 2010; 102(23): 1756-9. [http://dx.doi.org/10.1093/jnci/djq427] [PMID: 21044964]

[95] Sigman M. Introduction: Personalized medicine: what is it and what are the challenges? Fertil Steril 2018; 109(6): 944-5. [http://dx.doi.org/10.1016/j.fertnstert.2018.04.027] [PMID: 29935651]

[96] Dziąbowska K, Czaczyk E, Nidzworski D. Detection methods of human and animal influenza virus-current trends. Biosensors (Basel) 2018; 8(4): E94. [http://dx.doi.org/10.3390/bios8040094] [PMID: 30340339]

[97] Brendish NJ, Schiff HF, Clark TW. Point-of-care testing for respiratory viruses in adults: The current landscape and future potential. J Infect 2015; 71(5): 501-10. [http://dx.doi.org/10.1016/j.jinf.2015.07.008] [PMID: 26215335]

[98] Egilmezer E, Walker GJ, Bakthavathsalam P, *et al.* Systematic review of the impact of point-of-care testing for influenza on the outcomes of patients with acute respiratory tract infection. Rev Med Virol 2018; 28(5): e1995. [http://dx.doi.org/10.1002/rmv.1995] [PMID: 30101552]

[99] Berry L, Lansbury L, Gale L, Carroll AM, Lim WS. Point of care testing of Influenza A/B and RSV in an adult respiratory assessment unit is associated with improvement in isolation practices and reduction in hospital length of stay. J Med Microbiol 2020; 69(5): 697-704. [http://dx.doi.org/10.1099/jmm.0.001187] [PMID: 32250239]

CHAPTER 6

Impact of Influenza and COVID-19 Pandemics on Circulation of Seasonal Respiratory Viruses

Abstract: The emergence of novel respiratory viruses has often been accompanied by the disappearance of existing circulating strains. In this chapter, the impact of pandemics on the circulation of seasonal respiratory viruses is discussed. Some issues arising from the spread of pandemic viruses and underlying the choices of a strategy of fighting COVID-19 are under debate.

Keywords: COVID-19, Epidemics, Influenza, Pandemics, Viral respiratory infections.

INTRODUCTION

The first COVID-19 case happened in November 2019. At the end of December 2019, China reported an infection of unknown causes in Wuhan, Hubei province. Soon after, Chinese scientists identified a novel beta-coronavirus. The virus is now referred to as severe acute respiratory syndrome coronavirus 2 (SARS-CoV-2), and the disease is called COVID-19 (coronavirus disease 2019) [1]. SARS-CoV-2 properties meet three basic requirements that make the strain pandemic – it is new for the human's immune system, highly virulent for infected persons, and highly transmissible (see Chapter **5**, Fig. **1**). SARS-CoV-2 is less transmissible than measles but more contagious than influenza. Globally, since 31 December 2019 and as of 16 March 2021, there have been 119,960,700 confirmed cases of COVID-19, including 2,656,822 deaths, reported to WHO [2].

EMPTY SPACE GOES AGAINST THE LAWS OF NATURE

In nature, each of the various species, from higher mammals to microorganisms that constitute a community, occupies its ecological niche. At the end of the 20th century, the incidence of dangerous infectious diseases like smallpox, polio, or bubonic plague, which claimed hundreds of thousands of lives in the past, was

Irina Kiseleva and Natalie Larionova

significantly reduced. After defeating these infections, a large ecological niche was freed up. Nevertheless, "nature abhors a vacuum". Empty or unfilled spaces are unnatural as they go against the laws of nature. Therefore, the vacant niche began to be filled with other viruses.

Historically, the emergence of novel influenza viruses has often been accompanied by the disappearance of circulating seasonal strains. This phenomenon of crowding out seasonal influenza virus strains with new pandemic viruses is well known and described in numerous publications [3]. Novel pandemic influenza viruses enter the human population, displace previously circulating strains from the niche they occupy, and fill it. Palese and Wang [3] proposed that "the elimination of seasonal strains during virus pandemics is a process mediated, at the population level, by humoral immunity" in the human population.

The situation with COVID-19 looks different. This infection appeared unexpectedly, and there was not an empty spot for it; a unique niche was already occupied with other respiratory viruses. However, SARS-CoV-2 viruses filled out the same niches as influenza-like diseases did (Fig. **1**).

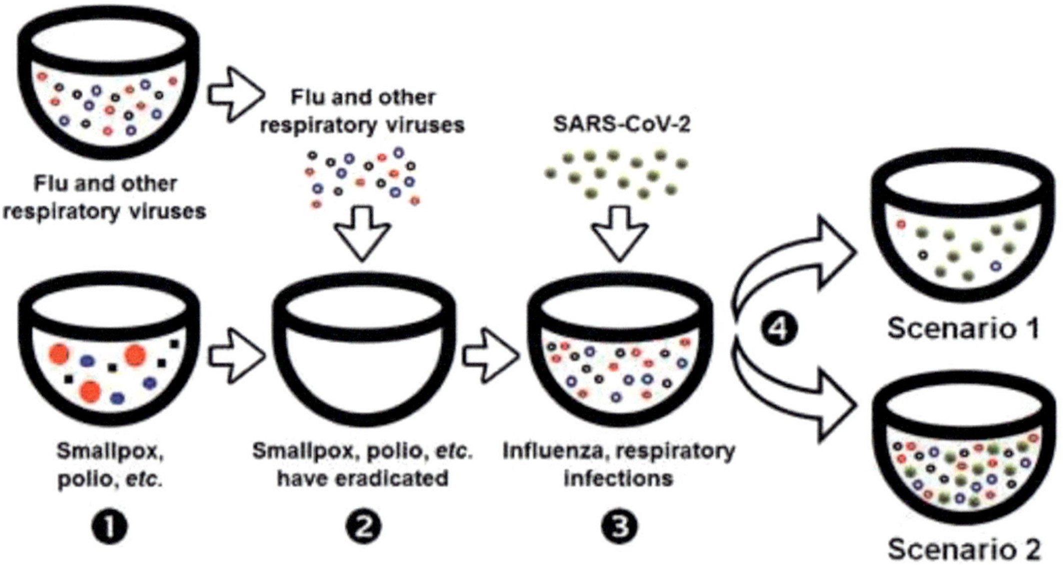

Fig. (1). "Natūra abhorret vacuum." The COVID-19 pandemic probably can develop in one of two scenarios: SARS-CoV-2 viruses will cocirculate with other respiratory viruses, or SARS-CoV-2 viruses will displace them.

Of course, many respiratory viruses, including influenza, have been circulated in humans before the eradication of smallpox or polio in the 1980s. In Fig. (**1**), we hypothesized how and what filled the empty spot vacated after defeating smallpox or polio.

What is happening today? It can be assumed that there are at least two scenarios for the development of subsequent events. Will SARS-CoV-2 being the most aggressive virus, displace other, more harmless strains (Fig. **1**, scenario **1**), or will they cocirculate together (Fig. **1**, scenario **2**)? As of today, it is uncertain.

WARS, STRESS AND INFECTIOUS DISEASES

In this paragraph, we would like to take a short break from COVID-19 and remind readers how infectious diseases, including respiratory infections, proceeded during devastating wars. Down the ages, people fought one against another; for centuries, wartime epidemics of infectious diseases have decimated the fighting strength of armies and brought havoc to the civil populations. Before the 20th century, deaths from diseases regularly exceeded those caused by military actions. For instance, during French Revolution and Napoleonic Wars (1792–1815), the mortality ratio (000's) [a measure of the number of deaths due to infectious diseases/the number of deaths from battle and wounds] was 7.6. On the contrary, from the early 20th century, during the Russo-Japanese War (1904–1905), World War I (1914–1918), and then World War II (1939–1945), the mortality ratio decreased more than twentyfold [4]. They were the wars in history in which more people were killed by military action than by infectious diseases [5]. During 1990-1991, in Operations Desert Shield and Desert Storm, infectious diseases were not a major cause of lost manpower, in contrast to previous wars involving the US military personnel [6].

Therefore, the early years of the 20th century marked a watershed in the epidemiologic history of wars. The beginning of the 20th century was marked by improvements in disease control and military hygiene and signaled the start of an enduring temporal trend in which more soldiers died in battle than because of infectious diseases. By the outbreak of World War II, the most dangerous and epidemiologically important pathogens had been identified, and advances in vaccination and chemotherapy had contributed significantly to tremendous efforts in preventive medicine, treatment, and the creation of a comprehensive health infrastructure [4, 6].

Nevertheless, globally, wars, revolutions, and coups have typically been accompanied by a pronounced increase in the occurrence of infectious diseases. Similarly, in civil populations, wars of the late 20th and early 21st centuries have continued to be associated with a marked increase in epidemic activity [4]. One of the most striking examples is the Spanish influenza pandemic. The 1918 virus spread globally to all continents with the mobilization of the military troops through the massive movement of people associated with World War I [7, 8].

Besides, long-term health consequences for military operations were described. A consistent and significant pattern of the increasing prevalence of chronic obstructive pulmonary disease and asthma in Iraq and Afghanistan veterans between 2003 and 2011 was found [9].

An interesting fact was noticed in the former USSR during the Great Patriotic War of 1941-1945, which claimed about 27 million lives of Soviet citizens. Besides deaths from battle and wounds, people died of hunger, heart attacks, and various diseases resulted in 2 million losses. Nevertheless, acute respiratory viral infections were very rare in wartime. People who got sick a lot in peacetime did not catch a cold in wartime usually. During stress in wartime, the human body was not susceptible to many infections. Some believe that viral diseases are directly related to the human psyche [10], therefore, at the time of intense stress, the brain "forbids" the body to get sick, and it is mobilized for defensive action (so-called "fight-or-flight" response) [11]. During the hostilities, the soldiers had to spend hours in the cold or even ice water and sit in trenches, but they did not get acute respiratory infections.

On the contrary, the stress experienced by people in peacetime has a different origin and provokes the appearance of diseases [12], including infectious diseases. When stressed people are crowded together, as residents of refugee camps, prisoners, students, and hospitalized people, their proximity and compromised immune system provide a great opportunity for a pathogen to enhance its virulence [13].

A relationship between psychological stress and parameters of the immune system in humans is well described [14 - 18]. So-called acute, time-limited stressors that lasted between 5 and 100 min reliably affect the immune system, including an increase in immune parameters, especially natural immunity. In contrast, chronic stressors do have negative effects on almost all functional measures of the immune system. It means that acute stressors with the temporal parameters of the "fight-or-flight" situations elicite potentially beneficial changes in the immune system and *vice versa*; the more a stressor becoming chronic, the more components of the immune system are affected in a potentially detrimental way [19].

IMPACT OF INFLUENZA PANDEMICS ON CIRCULATION OF SEASONAL RESPIRATORY VIRUSES

Wars, stresses, and prolonged force majeure situations upset the delicate balance in the host-pathogen interactions. Nevertheless, not only wars but also global outbreaks are stressful for society. What happens to other infectious diseases during the time of global disasters such as pandemics?

Numerous viruses are associated with human respiratory infections. To better understand the threat that these viruses pose to humankind, it is often valuable to study newly identified viruses, not in isolation, but in comparison with other related viruses [20]. In this paragraph, we briefly discuss relationships between newly emerged pandemic viruses and circulating seasonal strains.

The first virologically documented pandemic of Spanish influenza caused by the influenza A (H1N1) virus occurred in 1918–1919 [21]. According to various estimates, this unprecedented severe pandemic killed 40 to 50 million people and caused a huge number of severe forms of the disease. In subsequent years, the pathogen of the Spanish influenza A (H1N1) moved into the category of seasonal pathogens.

Unfortunately, we were unable to find available literature on the prevalence of other respiratory infections during rampages of the most destructive 1918 influenza pandemic. But a huge range of publications describing the spread of common respiratory infections during the next 1957, 1968, and 2009 influenza pandemics are available.

The evolution of influenza viruses has been discussed in detail in Chapter **3**. Nevertheless, we will briefly discuss some of its aspects again. In 1957, a new pandemic began, caused by the Asian influenza A (H2N2) virus, killing 1.5 million people. After the pandemic ended, Asian influenza viruses circulated for 11 years, causing annual epidemics.

A new influenza A (H3N2) pandemic virus, called Hong Kong flu, replaced the Asian influenza viruses in 1968. It killed 1 million people. Drift variants of the influenza A/Hong Kong (H3N2) virus continue to circulate to date.

In 1977, the subtype A (H1N1) virus came back into circulation, causing morbidity among young, immunologically naive people. Studies have shown that this virus was genetically close to the A (H1N1) virus circulating in the 1950s.

It is noteworthy that the emergence of each new pandemic virus was regularly marked by the disappearance of its pandemic predecessor, drifting in seasonal epidemics. There is an elegant explanation of the mysterious phenomenon of the displacement of a previously circulating virus by a new pandemic virus, according to which the mechanism leading to the elimination of the previous pathogen involves an increase in the production of human antibodies to the conserved stem of viral hemagglutinin protein. These antibodies are enough to displace an obsolete virus [3, 22]. An exception was the joint circulation of A (H3N2) and A (H1N1) viruses after 1977 when A (H1N1) viruses returned to the community and were unable to displace A (H3N2) viruses. Both subtypes alternately or simultaneously circulated among people until 2009.

In 2009, the serotype A (H1N1) viruses drifted since 1977 were supplanted by a completely new virus of the same serotype A (H1N1)pdm09, which caused a new moderate pandemic [23, 24]. Afterward, the virus A (H1N1)pdm09 has moved into the category of seasonal viruses and continued to circulate with viruses A (H3N2) until now.

What is happening to seasonal respiratory viruses in the interpandemic and pandemic periods? A vivid example of this is the situation covering the period before and after the last influenza pandemic of 2009 (2007-2012). In sum, influenza H1N1pdm09 pandemic viruses had distinct effects on the circulation of other respiratory viruses when they first appeared. On the contrary, no effect was seen when novel H1N1pdm09 strains evolved from being a pandemic to a seasonal virus. Pandemic influenza virus strains replaced circulating seasonal influenza virus strains; seasonal influenza H3N2 and B infections were hardly detected during the pandemic. However, the important thing is that the circulation of other respiratory pathogens remained unchanged [25].

Globally, seasonal influenza demonstrated wide variations in incidence depending on the epidemic season and the antigenic properties of the circulating strains. In contrast, yearly incidence rates for such non-influenza pathogens as a respiratory syncytial virus (RS-virus) are fairly stable. At the same time, the detection of the influenza virus had always been significantly below the RS-virus in influenza seasons before 2009. The 2009 pandemic significantly altered the detection rates of respiratory viruses, establishing influenza viruses as the major respiratory pathogens worldwide [26]. For instance, a study on the circulation of respiratory viruses other than influenza during the 2009 pandemic in Spain revealed that apart from influenza H1N1pdm09 viruses, other respiratory viruses were detected in 39.9% of patients. The co-infection of two or more respiratory viruses was observed in approximately 15% of cases. The highest detection rates were obtained for human rhinovirus and RS-virus, representing 43.7% and 31.4% of positive non-influenza samples, correspondingly [26].

It has been hypothesized that, at least in some European countries, a viral interference of human rhinovirus could have delayed the epidemic evolution of pandemic influenza H1N1pdm09 virus in the fall of 2009. Similarly, the influenza H1N1pdm09 virus could interfere with another respiratory pathogen, the RS-virus, in the same period [26].

Some authors have postulated that a possible interference of other respiratory viruses such as human rhinovirus and RS-virus with influenza H1N1pdm09 virus during the pandemic could change the epidemic wave of pandemic influenza [27 - 29].

IMPACT OF THE FIRST COVID-19 PANDEMIC ON CIRCULATION OF SEASONAL RESPIRATORY VIRUSES AND OTHER PATHOGENS

Humans may be constantly infected with microorganisms resident in their airways and by inhaled aerosols [30]. A single infection is caused by the simultaneous synergistic interaction of a variety of pathogens. In the context of COVID-19, SARS-CoV-2-infected individuals may be co-infected with **(i)** influenza virus, **(ii)** influenza-like respiratory virus, **(iii)** and other pathogens such as non-respiratory viruses, *Mycobacterium tuberculosis*, fungi, mycoplasma, *etc*. (Fig. **2**). Each of these variants may negatively influence COVID-19 treatment. Important questions: Should we expect an increase or decrease in the number of coinfections between SARS-CoV-2 and other respiratory pathogens? and what do we observe now?

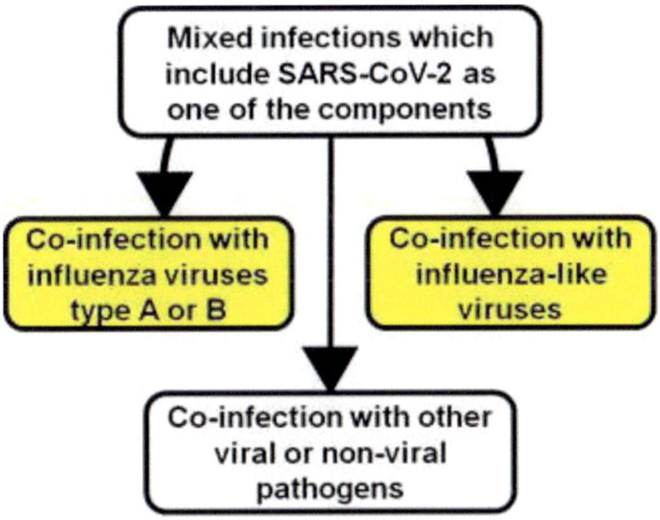

Fig. (2). Mixed infections of SARS-CoV-2 with other pathogens in COVID-19 pandemic [31]. SARS-CoV-2-infected individuals may be co-infected with influenza or influenza-like respiratory virus or with other pathogens.

COVID-19 can be mixed with other viral infections. Some of these co-infections can be effectively treated, while only supportive treatment is the mainstay of treatment for others [32]. Most publications, which describe co-occurred infections in which COVID-19 is involved, focus on influenza and influenza-like diseases. There are only a few papers on other infections. In this chapter, we have strictly confined ourselves to influenza and other viral respiratory infections.

Co-Infection of SARS-CoV-2 with Human Influenza Viruses

During the epidemic of SARS-CoV-2 in Beijing (China), influenza viruses accounted for a large proportion of respiratory virus infection [33], wherein a recent study revealed that Wuhan citizens with documented mixed infection of SARS-CoV-2 and influenza A or B did not appear to show a more severe clinical condition; they displayed similar clinical characteristics as those patients with SARS-CoV-2 infection only [34].

Some believe that a global co-infection of SARS-CoV-2 with the influenza virus is rare [35]. There have been a limited number of COVID-19-influenza co-infection cases reported from different countries worldwide; the USA, influenza A [32, 36], Turkey, influenza A [35], Germany, influenza A [37], Iran, influenza A [38, 39], China, influenza A [34, 40, 41], Japan, influenza A [42], Spain, influenza A [43], Turkey, influenza B [35], China, influenza B [34], Spain, influenza B [43], and China, Spain, simultaneously influenza A & B [44], *etc.* A study performed in Istanbul, Turkey, revealed that only 6 (0.54%) out of 1103 COVID-19 patients were diagnosed co-infected with influenza [35].

In any situation, influenza preparedness measures are crucial, especially for high-risk groups; the influenza vaccination coverage rate has to be increased up to at least 70% [45]. COVID-19 measures may reduce the number of cases of influenza infection and another way around. Seasonal influenza cases in the Northern Hemisphere usually peak in February and tail off by the end of May. Unusually, this year lab-confirmed cases of influenza dropped dramatically in early April, soon after WHO declared the COVID-19 pandemic. It has been hypothesized that the duration of the influenza season was shortened, and influenza incidence was substantially reduced by control measures aimed at interrupting SARS-CoV-2 transmission [46].

Co-Infection of SARS-CoV-2 with Influenza-Like Viruses

Besides influenza viruses, other respiratory viruses occupy a large place in the epidemiological process. Recently, influenza-like viruses show different levels of mixed infections with SARS-CoV-2 [47]. In light of the foregoing, the COVID-19 pandemic probably can develop in one of two scenarios: **(i)** SARS-CoV-2 viruses will cocirculate with other respiratory viruses; and **(ii)** SARS-CoV-2 viruses will displace other viral pathogens (Fig. **1**).

The last evidence suggests that unlike influenza viruses that demonstrated some reduction in activity, the co-infection rates between SARS-CoV-2 and other respiratory pathogens are higher than expected [48]. Such co-infection can aggravate the course of the disease and represent a challenge for the diagnosis and treatment. The study, performed in California (the US), revealed that out of all patients with confirmed SARS-CoV-2 infection, 20.7% were positive for one or more other pathogens; the most common were RS-virus, enterovirus, rhinovirus, and seasonal coronavirus [49]. In the study performed in China, among 32 confirmed COVID-19 cases, 14 patients were infected with other pathogens, including viruses, bacteria, and fungi. Only four persons were infected with respiratory viruses – RS-virus, human parainfluenza virus, human metapneumovirus, and rhinovirus [50].

During the beginning of COVID-19 spread in Japan, influenza and human metapneumovirus were prevalent among children (the study was performed in the Furano region). Three children were diagnosed with COVID-19; two of them were co-infected with influenza and human metapneumovirus [51]. Some believed that co-infection with two respiratory pathogens is uncommon in adults and mainly inherent in children [52]. Nonetheless, mixed infection of metapneumovirus and SARS-CoV-2 was detected in an adult patient in providence (RI, USA) [52].

In sum, among concomitant to SARS-CoV-2 pathogens, RS-virus and rhinovirus prevailed and were the most frequent causes of acute respiratory diseases with especially severe diseases in young children and the elderly. Primary RS-virus infections may lead to severe bronchiolitis and pneumonia [53, 54]. Rhinovirus usually causes rhinopharyngitis but may also cause illness of the lower respiratory tract (LRT) and asthma exacerbations and has been associated with severe acute LRT infections [53, 54]. The post-infection immunity to these infections is short-lasting and leads to the repeated circulation of these viruses in the same population. The co-infection of such seasonal pathogens, along with influenza viruses, with new COVID-19 pandemic can aggravate the course of the disease and represents a challenge for the diagnosis and treatment [45, 55, 56].

Co-Infection of SARS-CoV-2 with Other Viral and Non-Viral Pathogens

Current evidence does not support a high rate of bacterial co-infections in patients with COVID-2 [57]. On the contrary, patients have a low proportion of COVID-19 bacterial co-infection [47]. Bengoechea and Bamford [1] noticed that the limited reference to non-viral co-infections in COVID-19 is surprising. Only a few publications describing co-infections with non-viral pathogens are available in PubMed biomedical literature database. In particular, limited information regarding mixed infections of SARS-CoV-2 with bacterial and fungal infectious agents, including *Mycoplasma pneumonia* [58, 59], *Mycobacterium tuberculosis* [60 - 62], *Legionella pneumophila* [63], and *Aspergillus fumigatus* [64, 65], *etc.* could be found. Non-respiratory viral pathogens such as cytomegalovirus [66] or HIV [67] also do not play a crucial role in viral-viral mixed infections. Therefore, there is a paucity of data to support the association of SARS-CoV-2 and these pathogens with the severity of co-infection.

Impact of SARS-CoV-2 on Circulation of Seasonal Influenza and Other Respiratory Viruses

Virus-virus interactions are known for a long time but not completely studied yet [68, 69]. There is little information about the mechanisms of their interference and why some viruses displace others.

In the past, the appearance of each new influenza pandemic virus was regularly accompanied by the disappearance of circulating seasonal viruses. The same trend is observed today during the COVID-19 pandemic. Multiple studies reported that rates of influenza and a number of other viral respiratory diseases such as parainfluenza virus, adenovirus, metapneumovirus, and bocavirus infections had fallen significantly during recent coronavirus lockdowns [36, 46, 70, 71]. According to WHO, as of the information of 1 March 2021, global influenza activity is at lower levels than expected for this time of the year despite continued or even increased testing for influenza in some countries. In the Northern Hemisphere, influenza activity remained below baseline, while in the Southern Hemisphere, influenza activity was reported at inter-seasonal level [72]. In contrast, respiratory syncytial virus and rhinovirus are the most commonly identified viruses, followed by SARS-CoV-2 [73].

WHO warns that its information should be interpreted with caution because of possible influences of both COVID-19 activity and distance measures. "Despite the ongoing global response to the COVID-19 pandemic, we cannot lose sight of other significant public health issues, including influenza. Influenza affects every country every year and takes its deadly toll. As we enter the Southern Hemisphere influenza season and begin planning for the Northern Hemisphere season, we must ensure that influenza remains a top priority," Dr. Ghebreyesus, WHO Director-General, said at the WHO press-conference on 15 June 2020 [74]. Dr. Ghebreyesus noticed that today we faced a dramatic decrease in the number of specimens tested for influenza globally. These disruptions may have short and long-term effects, such as loss of capacities to detect and report new influenza viruses with pandemic potential.

Thus, SARS-CoV-2 may significantly affect infection with other respiratory viruses. However, it is difficult to distinguish its direct impact on reducing respiratory infections from the effect of declining Influenza surveillance because of the COVID-19 pandemic.

LESSONS LEARNED FROM THE 1918 GREAT INFLUENZA PANDEMIC

The great Spanish influenza pandemic represents a prime example of a worst-case scenario for such global disease outbreaks like COVID-19. Franchini *et al.* [75] suggested that "it is not just assonances of SARS-CoV-2 with the 1918-1919 flu pandemic, but an absolute coincidence." During the Spanish pandemic, the mortality rate of the susceptible population was about 1-2%, similar to the mortality rate of COVID-19. However, the former's death rate of 2% of the total population translates into 150 million deaths today, but in reality, since 31 December 2019 and as of 16 March 2021, the spread of the new coronavirus has led to the death rate of 2.2% reported to WHO [2]. At this point, the probability that COVID-19 reaches anything close to the Great Influenza pandemic seems remote, especially if taking into consideration the epidemiological differences, mitigating measures, and advances in science and public health.

As of 16 March 2021, mortality rate (observed case-mortality ratio) in the most affected country, Mexico, was 9.0%. However, the mortality rate in the majority of countries worldwide did not exceed 4.1% [2].

Pennington [5], talking about lessons learned from past pandemics, suggested, "Despite many decades of research, our predictive ability is still little better than those who went to war in 1914 expecting it to be over by Christmas."

History repeats itself, and it is important to learn from the past. Like the current COVID-19 pandemic, the 1918 Spanish flu began from a zoonotic transmission, progressed rapidly, and lead to death primarily by pneumonia in severe cases [76, 77]. Like the 1918 Spanish pandemic, the current COVID-19 pandemic disrupted the daily life of the plagued countries [78, 79].

In 2010, Morens *et al.* [80] suggested that acute respiratory distress syndrome may be a common endpoint for 1918 Spanish flu and the two viral pulmonary diseases thought to be associated with cytokine storms, severe acute respiratory syndrome, and severe hantavirus pulmonary syndrome. Today, 10 years later, we can extend this suggestion to COVID-19.

In general, all measures and strategies adopted in 2020 were the same in 1918-1919, with the same early loosening of control. The major problems, such as, for instance, social distancing, intrafamily spread, personal protective equipment, the need for vaccination, *etc*., had already been detected by the time. Unfortunately, despite all efforts in studying COVID-19, little progress in controlling this infection has been made so far.

One hundred years ago, there was no available vaccine against 1918 Spanish flu. In contrast, an unprecedented effort in developing the anti-COVID-19 vaccine is ongoing. It is strongly hoped that this effort will be successful. Over two hundred teams all over the world are working hard with a diverse set of strategies to develop a potent and harmless vaccine against COVID-19. As of 16 March 2021, 182 vaccines are in preclinical evaluation, 82 are in phases 1, 2, or 3 clinical trials [81]. Most new candidate vaccines go through a few years of testing before they appeared on the market. Therefore, scientists put forward a wide variety of theories on the available immunobiological preparations, which may be used in the fight against this disease.

CAN VACCINES AGAINST OTHER INFECTIONS PROTECT US AGAINST COVID-19 OR MINIMIZE ITS SEVERITY?

Since the COVID-19 outbreak began, the role of vaccination with a live attenuated Bacille Calmette-Guérin vaccine (BCG) against tuberculosis in preventing novel coronavirus infection has been discussed in numerous publications [82 - 85]. Based on the hypothesis that BCG can enhance the reactivity of the innate immune system, some have speculated that the BCG vaccine may be used as a preventive or therapeutic measure against COVID-19. Nonetheless, the results were conflicting, and later studies did not confirm a correlation between COVID-19 occurrence and BCG vaccination rates. After a detailed analysis of all available data, the WHO made the official conclusion that there is no evidence that the BCG vaccine protects people against infection with the COVID-19 virus [86]. Other researchers also concluded that there is no reasonable evidence to recommend BCG vaccination for the prevention of COVID-19 [87, 88].

In general, a vaccine against infectious disease aims to protect a target against illness, not necessarily infection itself [89]. Some authors suggest that the first vaccines against COVID-19 may prevent people from getting sick or dying but will not stop them from catching the coronavirus. Taking into account this fact, we have to be warned that some coronavirus vaccinations might be more harmful than helpful for those who are desperate for protection. Some caution is prudent, and the gains that have been recently made in understanding the interactions between the host immune system and viruses should be fully exploited [90].

A coronavirus vaccination may lead to negative consequences, including the disease being enhanced by the negative effect of the infection in a person instead of defeated [91]. The "cytokine storm," an uncontrolled over-production of soluble markers of inflammation, appears to be one of the most dangerous and potentially life-threatening events related to COVID-19 [92]. Infection with circulating SARS-CoV-2 virus followed by vaccination may induce severe disease [93]. Potential risks associated with the vaccine against COVID-19 are described in several publications [91, 93 - 97]. The authors suggest that vaccination could make subsequent SARS-CoV-2 infection more severe. There are two major syndromes described for vaccine-enhanced disease, antibody-dependent enhancement and vaccine-associated enhanced respiratory disease. This already has happened before with vaccines against feline coronavirus [91].

Homology between human and viral proteins is an established factor in viral- or vaccine-induced autoimmunity [95, 97]. Immunological priming could involve autoimmunity in lung tissues due to previous exposure to the coronavirus spike protein. Thus, anti-COVID-19 vaccine developers have to prove whether their vaccines cause the same types of immune system malfunctions that have been observed in the past.

We have to remember that influenza remains an ever-present cause of disease and death globally [98]. Patients co-infected with SARS-CoV-2 and influenza virus have a higher risk of poor outcomes. Given the potential for influenza and SARS-CoV-2 co-infections globally [35, 99], especially during the flu season, it has been suggested that flu vaccination could be used to indirectly control COVID-19 [37, 45, 100]. Of course, a flu vaccine will not fully protect against coronavirus infection, but vaccination with a proven and safe influenza vaccine, especially a live one, will affect the incidence of influenza – SARS-CoV-2 mixed infections and thus reduce the severity of coronavirus disease.

Some authors suggested that in the fall the influenza vaccine for larger groups of the population should be recommended to simplify clinicians' work when the next wave of COVID-19 is expected in parallel with the start of the influenza season [35, 101]. Other researchers were not optimistic [94, 102, 103]. In any case, understanding the relationship between COVID-19 and influenza is very important.

Salem and El-Hennawy [100] hypothesized that immunity against influenza infection may foster immunity against SARS-CoV-2. This suggestion is supported by studies showing cross-reactivity of immunity between influenza virus and coronavirus due to the similarity in their structures. Influenza vaccination itself would also generate sustained immunity that may enhance immunity against

SARS-CoV-2. The authors suggest that the immunity induced by the influenza virus has more beneficial effects for COVID-19 than those suggested for BCG or MMR (measles, mumps, and rubella) vaccines. More than 500 million people are vaccinated against the flu every year based on recommendations from WHO about the composition of influenza virus vaccines [74, 104]. According to a study [100], given the high safety of the influenza vaccine confirmed by decades of vaccination of millions of vaccinees, influenza vaccines can be used to minimize the severity of COVID-19 disease.

CONCLUDING REMARKS

With a wide diversity of zoonotic viruses capable of overcoming host range restrictions and acquiring the ability to spread in an immunologically naive population, the emergence of a new potentially pandemic virus in the human population is always on the horizon [20].

The pandemic invasion of a new SARS-CoV-2 virus in 2019 provided a unique possibility to learn quickly more about the pathogenesis of respiratory viruses. When faced with a new virus rapidly emerging in the human population, we have a limited knowledge base to work with. The clock is ticking, and we have to learn quickly about this new virus, simultaneously using our knowledge on previous influenza outbreaks, which have a lot in common with the recent COVID-19 outbreak [98, 105].

Having analyzed influenza viral ecology and relationships between animal and human influenza viruses, Ozawa and Kawaoka [106] suggested that "complete control of influenza viruses seems impossible." It looks like it is not futile to think that complete influenza eradication is possible in the nearest future. As for today, the COVID-19 pandemic is far from over, and we, like with influenza, are far from defeating it. Meanwhile, nobody can predict anything. We are just at the beginning of a long and winding road.

ACKNOWLEDGEMENTS

We wish to thank Dr. Elena Grigorieva and Dr. Mohammad Al Farroukh for their editorial assistance, critical reading, and help in the collection of scientific information.

REFERENCES

[1] Bengoechea JA, Bamford CGG. SARS-CoV-2, bacterial co-infections, and AMR: the deadly trio in COVID-19? EMBO Mol Med 2020; 12(7): e12560. [http://dx.doi.org/10.15252/emmm.202012560] [PMID: 32453917]

[2] WHO. Coronavirus disease (COVID-19) dashboard. 2020. [Available from: https://covid19.who.int/]

[3] Palese P, Wang TT. Why do influenza virus subtypes die out? A hypothesis. MBio 2011; 2(5): e00150-11.
[http://dx.doi.org/10.1128/mBio.00150-11] [PMID: 21878571]

[4] Smallman-Raynor MR, Cliff AD. Impact of infectious diseases on war. Infect Dis Clin North Am 2004; 18(2): 341-68.
[http://dx.doi.org/10.1016/j.idc.2004.01.009] [PMID: 15145384]

[5] Pennington H. The impact of infectious disease in war time: a look back at WW1. Future Microbiol 2019; 14: 165-8.
[http://dx.doi.org/10.2217/fmb-2018-0323] [PMID: 30628481]

[6] Hyams KC, Hanson K, Wignall FS, Escamilla J, Oldfield EC III. The impact of infectious diseases on the health of U.S. troops deployed to the Persian Gulf during operations Desert Shield and Desert Storm. Clin Infect Dis 1995; 20(6): 1497-504.
[http://dx.doi.org/10.1093/clinids/20.6.1497] [PMID: 7548499]

[7] Akin L, Gözel MG. Understanding dynamics of pandemics. Turkish J Med Sci 2020; 50(Si-1): 515-9.
[http://dx.doi.org/10.3906/sag-2004-133] [PMID: 32299204]

[8] Johnson NP, Mueller J. Updating the accounts: global mortality of the 1918-1920 "Spanish" influenza pandemic. Bull Hist Med 2002; 76(1): 105-15.
[http://dx.doi.org/10.1353/bhm.2002.0022] [PMID: 11875246]

[9] Pugh MJ, Jaramillo CA, Leung KW, *et al.* Increasing prevalence of chronic lung disease in veterans of the wars in Iraq and Afghanistan. Mil Med 2016; 181(5): 476-81.
[http://dx.doi.org/10.7205/MILMED-D-15-00035] [PMID: 27136656]

[10] Coughlin SS. Anxiety and depression: Linkages with viral diseases. Public Health Rev 2012; 34(2): 92.
[http://dx.doi.org/10.1007/BF03391675] [PMID: 25264396]

[11] Dhabhar FS, Malarkey WB, Neri E, McEwen BS. Stress-induced redistribution of immune cells--from barracks to boulevards to battlefields: a tale of three hormones--Curt Richter Award winner. Psychoneuroendocrinology 2012; 37(9): 1345-68.
[http://dx.doi.org/10.1016/j.psyneuen.2012.05.008] [PMID: 22727761]

[12] Lutsenko N. Why people were not infected with respiratory viruses during the Great Patriotic War? 2020. [Available from: https://nevnov.ru/799062-vrachi-dali-otvet-pochemu-vo-vremya-vov-lyudi-ne-boleli-orvi]

[13] GrrlScientist. Influenza: How the Great War helped create the greatest pandemic ever known. The Guardian 30 July 2014. Sect. Science. [Available from: https://www.theguardian.com/science/grrlscientist/2014/jul/30/influenza-pandemic-1918-viruses-biology-medicine-history]

[14] Solomon GF, Amkraut AA, Kasper P. Immunity, emotions and stress with special reference to the mechanisms of stress effects on the immune system. Psychother Psychosom 1974; 23(1-6): 209-17.
[http://dx.doi.org/10.1159/000286644] [PMID: 4153377]

[15] Udelman DL. Stress and immunity. Psychother Psychosom 1982; 37(3): 176-84.
[http://dx.doi.org/10.1159/000287570] [PMID: 6983697]

[16] Solomon GF, Segerstrom SC, Grohr P, Kemeny M, Fahey J. Shaking up immunity: psychological and immunologic changes after a natural disaster. Psychosom Med 1997; 59(2): 114-27.
[http://dx.doi.org/10.1097/00006842-199703000-00002] [PMID: 9088047]

[17] Sheridan JF, Dobbs C, Jung J, *et al.* Stress-induced neuroendocrine modulation of viral pathogenesis and immunity. Ann N Y Acad Sci 1998; 840: 803-8.
[http://dx.doi.org/10.1111/j.1749-6632.1998.tb09618.x] [PMID: 9629306]

[18] Cohen S, Janicki-Deverts D, Doyle WJ. Self-rated health in healthy adults and susceptibility to the common cold. Psychosom Med 2015; 77(9): 959-68.

[http://dx.doi.org/10.1097/PSY.0000000000000232] [PMID: 26397938]

[19] Segerstrom SC, Miller GE. Psychological stress and the human immune system: a meta-analytic study of 30 years of inquiry. Psychol Bull 2004; 130(4): 601-30. [http://dx.doi.org/10.1037/0033-2909.130.4.601] [PMID: 15250815]

[20] Belser JA. Assessment of SARS-CoV-2 replication in the context of other respiratory viruses. Lancet Respir Med 2020; 8(7): 651-2. [http://dx.doi.org/10.1016/S2213-2600(20)30227-7] [PMID: 32386570]

[21] Neumann G, Noda T, Kawaoka Y. Emergence and pandemic potential of swine-origin H1N1 influenza virus. Nature 2009; 459(7249): 931-9. [http://dx.doi.org/10.1038/nature08157] [PMID: 19525932]

[22] Krammer F, Palese P. Influenza virus hemagglutinin stalk-based antibodies and vaccines. Curr Opin Virol 2013; 3(5): 521-30. [http://dx.doi.org/10.1016/j.coviro.2013.07.007] [PMID: 23978327]

[23] Dawood FS, Iuliano AD, Reed C, *et al.* Estimated global mortality associated with the first 12 months of 2009 pandemic influenza A H1N1 virus circulation: a modelling study. Lancet Infect Dis 2012; 12(9): 687-95. [http://dx.doi.org/10.1016/S1473-3099(12)70121-4] [PMID: 22738893]

[24] Simonsen L, Spreeuwenberg P, Lustig R, *et al.* Global mortality estimates for the 2009 Influenza Pandemic from the GLaMOR project: a modeling study. PLoS Med 2013; 10(11): e1001558. [http://dx.doi.org/10.1371/journal.pmed.1001558] [PMID: 24302890]

[25] Meningher T, Hindiyeh M, Regev L, Sherbany H, Mendelson E, Mandelboim M. Relationships between A(H1N1)pdm09 influenza infection and infections with other respiratory viruses. Influenza Other Respir Viruses 2014; 8(4): 422-30. [http://dx.doi.org/10.1111/irv.12249] [PMID: 24698156]

[26] Navarro-Marí JM, Pérez-Ruiz M, Galán Montemayor JC, *et al.* Circulation of other respiratory viruses and viral co-infection during the 2009 pandemic influenza. Enferm Infecc Microbiol Clin 2012; 30 (Suppl. 4): 25-31. [http://dx.doi.org/10.1016/S0213-005X(12)70101-5] [PMID: 23116789]

[27] Casalegno JS, Ottmann M, Bouscambert-Duchamp M, Valette M, Morfin F, Lina B. Impact of the 2009 influenza A(H1N1) pandemic wave on the pattern of hibernal respiratory virus epidemics, France, 2009. Clin Microbiol Infect 2010; 16: 326-9. [http://dx.doi.org/10.1111/j.1469-0691.2010.03167.x] [PMID: 20158981]

[28] Mak GC, Wong AH, Ho WY, Lim W. The impact of pandemic influenza A (H1N1) 2009 on the circulation of respiratory viruses 2009-2011. Influenza Other Respir Viruses 2012; 6(3): e6-e10. [http://dx.doi.org/10.1111/j.1750-2659.2011.00323.x] [PMID: 22212717]

[29] Ånestad G, Nordbø SA. Virus interference. Did rhinoviruses activity hamper the progress of the 2009 influenza A (H1N1) pandemic in Norway? Med Hypotheses 2011; 77(6): 1132-4. [http://dx.doi.org/10.1016/j.mehy.2011.09.021] [PMID: 21975051]

[30] Babiuk LA, Lawman MJ, Ohmann HB. Viral-bacterial synergistic interaction in respiratory disease. Adv Virus Res 1988; 35: 219-49. [http://dx.doi.org/10.1016/S0065-3527(08)60713-7] [PMID: 3148270]

[31] Kiseleva I, Grigorieva E, Larionova N, Al Farroukh M, Rudenko L. COVID-19 in light of seasonal respiratory infections. Biology (Basel) 2020; 9(9): E240. [http://dx.doi.org/10.3390/biology9090240] [PMID: 32825427]

[32] Konala VM, Adapa S, Gayam V, *et al.* Co-infection with influenza A and COVID-19. Eur J Case Rep Intern Med 2020; 7(5): 001656. [http://dx.doi.org/10.12890/2020_001656] [PMID: 32399452]

[33] Li Y, Wang J, Wang C, *et al.* Characteristics of respiratory virus infection during the outbreak of 2019

novel coronavirus in Beijing. Int J Infect Dis 2020; 96: 266-9. [http://dx.doi.org/10.1016/j.ijid.2020.05.008] [PMID: 32389850]

[34] Ding Q, Lu P, Fan Y, Xia Y, Liu M. The clinical characteristics of pneumonia patients coinfected with 2019 novel coronavirus and influenza virus in Wuhan, China. J Med Virol 2020; 92(9): 1549-55. [http://dx.doi.org/10.1002/jmv.25781] [PMID: 32196707]

[35] Ozaras R, Cirpin R, Duran A, *et al.* Influenza and COVID-19 co-infection: Report of six cases and review of the literature. J Med Virol 2020; 92(11): 2657-65. [http://dx.doi.org/10.1002/jmv.26125] [PMID: 32497283]

[36] Nowak MD, Sordillo EM, Gitman MR, Paniz Mondolfi AE. Coinfection in SARS-CoV-2 infected patients: Where are influenza virus and rhinovirus/enterovirus? J Med Virol 2020; 92(10): 1699-700. [http://dx.doi.org/10.1002/jmv.25953] [PMID: 32352574]

[37] Wehl G, Laible M, Rauchenzauner M. Co-infection of SARS CoV-2 and influenza A in a pediatric patient in Germany. Klin Padiatr 2020; 232(4): 217-8. [http://dx.doi.org/10.1055/a-1163-7385] [PMID: 32392591]

[38] Hashemi SA, Safamanesh S, Ghafouri M, *et al.* Co-infection with COVID-19 and influenza A virus in two died patients with acute respiratory syndrome, Bojnurd, Iran. J Med Virol 2020; 92(11): 2319-21. [http://dx.doi.org/10.1002/jmv.26014] [PMID: 32410338]

[39] Khodamoradi Z, Moghadami M, Lotfi M. Co-infection of coronavirus disease 2019 and influenza A: A report from Iran. Arch Iran Med 2020; 23(4): 239-43. [http://dx.doi.org/10.34172/aim.2020.04] [PMID: 32271596]

[40] Wu D, Lu J, Liu Y, Zhang Z, Luo L. Positive effects of COVID-19 control measures on influenza prevention. Int J Infect Dis 2020; 95: 345-6. [http://dx.doi.org/10.1016/j.ijid.2020.04.009] [PMID: 32283283]

[41] Wu X, Cai Y, Huang X, *et al.* Co-infection with SARS-CoV-2 and influenza A virus in patient with pneumonia, China. Emerg Infect Dis 2020; 26(6): 1324-6. [http://dx.doi.org/10.3201/eid2606.200299] [PMID: 32160148]

[42] Azekawa S, Namkoong H, Mitamura K, Kawaoka Y, Saito F. Co-infection with SARS-CoV-2 and influenza A virus. IDCases 2020; 20: e00775. [http://dx.doi.org/10.1016/j.idcr.2020.e00775] [PMID: 32368495]

[43] Cuadrado-Payán E, Montagud-Marrahi E, Torres-Elorza M, *et al.* SARS-CoV-2 and influenza virus co-infection. Lancet 2020; 395(10236): e84. [http://dx.doi.org/10.1016/S0140-6736(20)31052-7] [PMID: 32423586]

[44] Wu Q, Xing Y, Shi L, *et al.* Coinfection and other clinical characteristics of COVID-19 in children. Pediatrics 2020; 146(1): e20200961. [http://dx.doi.org/10.1542/peds.2020-0961] [PMID: 32376725]

[45] Osterhaus ADME. Influenza and RSV in a COVID-19 world. The first ESWI webinar 11 June 2020. [Available from: https://eswi.org/eswi-tv/influenza-and-rsv-in-a-covid-19-world-full-webinar/]

[46] Jones N. How coronavirus lockdowns stopped flu in its tracks. Nature 21 May 2020. [http://dx.doi.org/10.1038/d41586-020-01538-8] [PMID: 32440021]

[47] Lansbury L, Lim B, Baskaran V, Lim WS. Co-infections in people with COVID-19: a systematic review and meta-analysis. J Infect 2020; 81(2): 266-75. [http://dx.doi.org/10.1016/j.jinf.2020.05.046] [PMID: 32473235]

[48] Sánchez-Duque JA, Orozco-Hernández JP, Marín-Medina DS, *et al.* Are we now observing an increasing number of coinfections between SARS-CoV-2 and other respiratory pathogens? J Med Virol 2020; 92(11): 2398-400. [http://dx.doi.org/10.1002/jmv.26089] [PMID: 32470211]

[49] Kim D, Quinn J, Pinsky B, Shah NH, Brown I. Rates of co-infection between SARS-CoV-2 and other

respiratory pathogens. JAMA 2020; 323(20): 2085-6.
[http://dx.doi.org/10.1001/jama.2020.6266] [PMID: 32293646]

[50] Li Z, Chen ZM, Chen LD, *et al.* Coinfection with SARS-CoV-2 and other respiratory pathogens in patients with COVID-19 in Guangzhou, China. J Med Virol 2020; 92(11): 2381-3.
[http://dx.doi.org/10.1002/jmv.26073] [PMID: 32462695]

[51] Kakuya F, Okubo H, Fujiyasu H, Wakabayashi I, Syouji M, Kinebuchi T. The first pediatric patients with coronavirus disease 2019 (COVID-19) in Japan; The risk of co-infection with other respiratory viruses. Jpn J Infect Dis 2020; 73(5): 377-80.
[http://dx.doi.org/10.7883/yoken.JJID.2020.181] [PMID: 32475878]

[52] Touzard-Romo F, Tapé C, Lonks JR. Co-infection with SARS-CoV-2 and human metapneumovirus. Rhode Isl Med J (2013) 2020; 103(2): 75-6.

[53] Costa LF, Queiróz DA, Lopes da Silveira H, *et al.* Human rhinovirus and disease severity in children. Pediatrics 2014; 133(2): e312-21.
[http://dx.doi.org/10.1542/peds.2013-2216] [PMID: 24420808]

[54] Li Y, Pillai P, Miyake F, Nair H. The role of viral co-infections in the severity of acute respiratory infections among children infected with respiratory syncytial virus (RSV): A systematic review and meta-analysis. J Glob Health 2020; 10(1): 010426.
[http://dx.doi.org/10.7189/jogh.10.010426] [PMID: 32566164]

[55] Osterhaus ADME. Clinical manifestations of COVID-19, influenza and RSV. The second ESWI webinar 18 June 2020. [Available from: https://eswi.org/eswi-tv/clinical-manifestations-of-covd-19-influenza-and-rsv-full-webinar/]

[56] Osterhaus ADME. Intervention strategies for COVID-19, influenza and RSV. The third ESWI webinar 23 June 2020. [Available from: https://player.vimeo.com/video/431944051]

[57] Rawson TM, Moore LSP, Zhu N, *et al.* Bacterial and fungal co-infection in individuals with coronavirus: A rapid review to support COVID-19 antimicrobial prescribing. Clin Infect Dis 2020; 71(9): 2459-68.
[http://dx.doi.org/10.1093/cid/ciaa530] [PMID: 32358954]

[58] Fan BE, Lim KGE, Chong VCL, Chan SSW, Ong KH, Kuperan P. COVID-19 and mycoplasma pneumoniae coinfection. Am J Hematol 2020; 95(6): 723-4.
[http://dx.doi.org/10.1002/ajh.25785] [PMID: 32173883]

[59] Gayam V, Konala VM, Naramala S, *et al.* Presenting characteristics, comorbidities, and outcomes of patients coinfected with COVID-19 and Mycoplasma pneumoniae in the USA. J Med Virol 2020; 92(10): 2181-7.
[http://dx.doi.org/10.1002/jmv.26026] [PMID: 32449972]

[60] Stochino C, Villa S, Zucchi P, Parravicini P, Gori A, Raviglione MC. Clinical characteristics of COVID-19 and active tuberculosis co-infection in an Italian reference hospital. Eur Respir J 2020; 56(1): 2001708.
[http://dx.doi.org/10.1183/13993003.01708-2020] [PMID: 32482787]

[61] Tadolini M, Codecasa LR, García-García JM, *et al.* Active tuberculosis, sequelae and COVID-19 co-infection: first cohort of 49 cases. Eur Respir J 2020; 56(1): 2001398.
[http://dx.doi.org/10.1183/13993003.01398-2020] [PMID: 32457198]

[62] Togun T, Kampmann B, Stoker NG, Lipman M. Anticipating the impact of the COVID-19 pandemic on TB patients and TB control programmes. Ann Clin Microbiol Antimicrob 2020; 19(1): 21.
[http://dx.doi.org/10.1186/s12941-020-00363-1] [PMID: 32446305]

[63] Arashiro T, Nakamura S, Asami T, *et al.* SARS-CoV-2 and Legionella co-infection in a person returning from a Nile cruise. J Travel Med 2020; 27(3): taaa053.
[http://dx.doi.org/10.1093/jtm/taaa053] [PMID: 32297939]

[64] van Arkel ALE, Rijpstra TA, Belderbos HNA, van Wijngaarden P, Verweij PE, Bentvelsen RG.

COVID-19 associated pulmonary aspergillosis. Am J Respir Crit Care Med 2020; 202(1): 132-5. [http://dx.doi.org/10.1164/rccm.202004-1038LE] [PMID: 32396381]

[65] Blaize M, Mayaux J, Nabet C, *et al.* Fatal invasive aspergillosis and coronavirus disease in an immunocompetent patient. Emerg Infect Dis 2020; 26(7): 1636-7. [http://dx.doi.org/10.3201/eid2607.201603] [PMID: 32343223]

[66] D'Ardes D, Boccatonda A, Schiavone C, *et al.* A case of coinfection with SARS-COV-2 and cytomegalovirus in the era of COVID-19. Eur J Case Rep Intern Med 2020; 7(5): 001652. [PMID: 32399451]

[67] Benkovic S, Kim M, Sin E. Four cases: HIV and SARS-CoV-2 Co-infection in patients from Long Island, New York. J Med Virol 2020; 92(11): 2338-40. [http://dx.doi.org/10.1002/jmv.26029] [PMID: 32427361]

[68] Henle W. Interference phenomena between animal viruses; a review. J Immunol 1950; 64(3): 203-36. [PMID: 15412251]

[69] Schultz-Cherry S. Viral interference: The case of influenza viruses. J Infect Dis 2015; 212(11): 1690-1. [http://dx.doi.org/10.1093/infdis/jiv261] [PMID: 25943204]

[70] Chan KH, Lee P-W, Chan CY, Lam KBH, Ho P-L. Monitoring respiratory infections in covid-19 epidemics. BMJ 2020; 369: m1628. [http://dx.doi.org/10.1136/bmj.m1628] [PMID: 32366507]

[71] Sullivan SG, Carlson S, Cheng AC, *et al.* Where has all the influenza gone? The impact of COVID-19 on the circulation of influenza and other respiratory viruses, Australia, March to September 2020. Euro Surveill 2020; 25(47): 2001847. [http://dx.doi.org/10.2807/1560-7917.ES.2020.25.47.2001847] [PMID: 33243355]

[72] WHO. Overview of influenza activity globally. Influenza Update N° 388 of 1 March 2021. [Available from: https://www.who.int/influenza/surveillance_monitoring/updates/latest_update_GIP_surveillance/en/]

[73] Calderaro A, De Conto F, Buttrini M, *et al.* Human respiratory viruses, including SARS-CoV-2, circulating in the winter season 2019-2020 in Parma, Northern Italy. Int J Infect Dis 2021; 102: 79-84. [http://dx.doi.org/10.1016/j.ijid.2020.09.1473] [PMID: 33017694]

[74] WHO. Coronavirus disease (COVID-19) pandemic. Press Conference of 15 June 2020 [Available from: https://www.who.int/emergencies/diseases/novel-coronavirus-2019]

[75] Franchini AF, Auxilia F, Galimberti PM, Piga MA, Castaldi S, Porro A. COVID 19 and Spanish flu pandemics: All it changes, nothing changes. Acta Biomed 2020; 91(2): 245-50. [http://dx.doi.org/10.23750/abm.v91i2.9625] [PMID: 32420960]

[76] Taubenberger JK, Kash JC, Morens DM. The 1918 influenza pandemic: 100 years of questions answered and unanswered. Sci Transl Med 2019; 11(502): eaau5485. [http://dx.doi.org/10.1126/scitranslmed.aau5485] [PMID: 31341062]

[77] Taubenberger JK, Morens DM. Influenza: the once and future pandemic. Public Health Rep 2010;125 Suppl 3(Suppl 3): 16-26. [http://dx.doi.org/10.1177/00333549101250S305] [PMID: 20568566]

[78] Trilla A, Trilla G, Daer C. The 1918 "Spanish flu" in Spain. Clin Infect Dis 2008; 47(5): 668-73. [http://dx.doi.org/10.1086/590567] [PMID: 18652556]

[79] Taubenberger JK, Morens DM. The 1918 Influenza pandemic and its legacy. Cold Spring Harb Perspect Med 2020; 10(10): a038695. [http://dx.doi.org/10.1101/cshperspect.a038695] [PMID: 31871232]

[80] Morens DM, Taubenberger JK, Harvey HA, Memoli MJ. The 1918 influenza pandemic: lessons for 2009 and the future. Crit Care Med 2010; 38(4) (Suppl.): e10-20.

[http://dx.doi.org/10.1097/CCM.0b013e3181ceb25b] [PMID: 20048675]

[81] WHO. Draft landscape of COVID-19 candidate vaccines. Update 10 August 2020. [Available from: https://www.who.int/publications/m/item/draft-landscape-of-covid-19-candidate-vaccines]

[82] Ozdemir C, Kucuksezer UC, Tamay ZU. Is BCG vaccination affecting the spread and severity of COVID-19? Allergy 2020; 75(7): 1824-7. [http://dx.doi.org/10.1111/all.14344] [PMID: 32330314]

[83] Curtis N, Sparrow A, Ghebreyesus TA, Netea MG. Considering BCG vaccination to reduce the impact of COVID-19. Lancet 2020; 395(10236): 1545-6. [http://dx.doi.org/10.1016/S0140-6736(20)31025-4] [PMID: 32359402]

[84] Meena J, Yadav A, Kumar J. BCG vaccination policy and protection against COVID-19. Indian J Pediatr 2020; 87(9): 749. [http://dx.doi.org/10.1007/s12098-020-03371-3] [PMID: 32519258]

[85] Macedo A, Febra C. Relation between BCG coverage rate and COVID-19 infection worldwide. Med Hypotheses 2020; 142: 109816. [http://dx.doi.org/10.1016/j.mehy.2020.109816] [PMID: 32408071]

[86] WHO. Bacille Calmette-Guérin (BCG) vaccination and COVID-19. Scientific Brief of 12 April 2020. [Available from: https://www.who.int/news-room/commentaries/detail/bacille-calmet-e-gu%C3%A9rin-(bcg)-vaccination-and-covid-19]

[87] Kumar J, Meena J. Demystifying BCG vaccine and COVID-19 relationship. Indian Pediatr 2020; 57(6): 588-9. [http://dx.doi.org/10.1007/s13312-020-1872-0] [PMID: 32358227]

[88] Riccò M, Gualerzi G, Ranzieri S, Bragazzi NL. Stop playing with data: there is no sound evidence that Bacille Calmette-Guérin may avoid SARS-CoV-2 infection (for now). Acta Biomed 2020; 91(2): 207-13. [http://dx.doi.org/10.23750/abm.v91i2.9700] [PMID: 32420947]

[89] Lauerman J, Paton J. The first COVID vaccines may not prevent COVID infection. Prognosis of 15 June 2020 [Available from: https://www.bloomberg.com/news/articles/2020-06-15/the-first-covid-vaccines-may-not-prevent-you-from-getting-covid]

[90] Burton DR, Walker LM. Rational vaccine design in the time of COVID-19. Cell Host Microbe 2020; 27(5): 695-8. [http://dx.doi.org/10.1016/j.chom.2020.04.022] [PMID: 32407707]

[91] Graham BS. Rapid COVID-19 vaccine development. Science 2020; 368(6494): 945-6. [http://dx.doi.org/10.1126/science.abb8923] [PMID: 32385100]

[92] Coperchini F, Chiovato L, Croce L, Magri F, Rotondi M. The cytokine storm in COVID-19: An overview of the involvement of the chemokine/chemokine-receptor system. Cytokine Growth Factor Rev 2020; 53: 25-32. [http://dx.doi.org/10.1016/j.cytogfr.2020.05.003] [PMID: 32446778]

[93] Eyal N, Lipsitch M, Smith PG. Human challenge studies to accelerate coronavirus vaccine licensure. J Infect Dis 2020; 221(11): 1752-6. [http://dx.doi.org/10.1093/infdis/jiaa152] [PMID: 32232474]

[94] Morris KV. The improbability of the rapid development of a vaccine for SARS-CoV-2. Mol Ther 2020; 28(7): 1548-9. [http://dx.doi.org/10.1016/j.ymthe.2020.06.005] [PMID: 32533920]

[95] Peeples L. News feature: avoiding pitfalls in the pursuit of a COVID-19 vaccine. Proc Natl Acad Sci USA 2020; 117(15): 8218-21. [http://dx.doi.org/10.1073/pnas.2005456117] [PMID: 32229574]

[96] Hotez PJ, Corry DB, Bottazzi ME. COVID-19 vaccine design: the Janus face of immune

enhancement. Nat Rev Immunol 2020; 20(6): 347-8. [http://dx.doi.org/10.1038/s41577-020-0323-4] [PMID: 32346094]

[97] Lyons-Weiler J. Pathogenic priming likely contributes to serious and critical illness and mortality in COVID-19 *via* autoimmunity. J Trans Autoimmun 2020; 3: 100051. [http://dx.doi.org/10.1016/j.jtauto.2020.100051] [PMID: 32292901]

[98] Dunning J, Thwaites RS, Openshaw PJM. Seasonal and pandemic influenza: 100 years of progress, still much to learn. Mucosal Immunol 2020; 13(4): 566-73. [http://dx.doi.org/10.1038/s41385-020-0287-5] [PMID: 32317736]

[99] Antony SJ, Almaghlouth NK, Heydemann EL. Are co-infections with COVID-19 and influenza low or underreported? An observational study examining current published literature including three new unpublished cases. J Med Virol 2020; 92(11): 2489-97. [http://dx.doi.org/10.1002/jmv.26167] [PMID: 32530531]

[100] Salem ML, El-Hennawy D. The possible beneficial adjuvant effect of influenza vaccine to minimize the severity of COVID-19. Med Hypotheses 2020; 140: 109752. [http://dx.doi.org/10.1016/j.mehy.2020.109752] [PMID: 32361099]

[101] Hanney SR, Wooding S, Sussex J, Grant J. From COVID-19 research to vaccine application: why might it take 17 months not 17 years and what are the wider lessons? Health Res Policy Syst 2020; 18(1): 61. [http://dx.doi.org/10.1186/s12961-020-00571-3] [PMID: 32513202]

[102] Mukherjee R. Global efforts on vaccines for COVID-19: Since, sooner or later, we all will catch the coronavirus. J Biosci 2020; 45(1): 68. [http://dx.doi.org/10.1007/s12038-020-00040-7] [PMID: 32385219]

[103] Merante D. SARS-CoV-2, from its current highly contagious spreading toward the global development of an effective and safe vaccine: challenges and uncertainties. Expert Opin Drug Saf 2020; 19(7): 771-4. [http://dx.doi.org/10.1080/14740338.2020.1773789] [PMID: 32442040]

[104] WHO. Recommendations on the composition of influenza virus vaccines. 2020. [Available from: https://www.who.int/influenza/vaccines/virus/recommendations/en/]

[105] Bogaert D, Dockrell DH. 100 years of influenza research seen through the lens of Covid-19. Mucosal Immunol 2020; 13(4): 561-2. [http://dx.doi.org/10.1038/s41385-020-0291-9] [PMID: 32358574]

[106] Ozawa M, Kawaoka Y. Cross talk between animal and human influenza viruses. Annu Rev Anim Biosci 2013; 1: 21-42. [http://dx.doi.org/10.1146/annurev-animal-031412-103733] [PMID: 25387011]

CHAPTER 7

Is it Possible to Eradicate Influenza?

Abstract: This short final chapter has been written in the form of a conclusion; it is devoted to summarizing the 100-year study of influenza. The main achievements in the study and fight against this infection are listed. The possibility of defeating influenza in the foreseeable future is discussed.

Keywords: Control for influenza, Influenza, Influenza eradication.

THE 18TH YEAR OF THE 20TH CENTURY

The 1918 Spanish influenza pandemic was a landmark in history and influenza research. The epidemiology, morbidity, and mortality of the devastating pandemic of 1918 have been described in many studies [1]. At the time of the most deadly influenza pandemic in history, at the disposal of physicians, there was nothing except symptomatic drugs. There were no specific immunobiological preparations, the pathogen was unknown, there was no adequate antiviral chemotherapy and interferons, and there was nothing to treat the deadly infection. Medical care for influenza was palliative. It was aimed to support the patients, but it was not an etiotropic treatment.

In those days, medical services recommended "open-air" treatment and taking an overdose of aspirin (acetylsalicylic acid) – 1.3 g per day *vs.* standard dose of 0.3 g per day (from the other sources – 30 g per day *vs.* the maximum allowable daily dose which is estimated to be 4 g per day) [2]. The treatment only aimed to control the symptoms. However, symptoms resembled all known diseases at once, and the physician did not know why and how to treat the patient.

ONE HUNDRED YEARS AFTER

Advances in Influenza Research

More than one hundred years have passed since the occurrence of the Spanish flu pandemic. Since the last eight decades after the first human influenza virus was

Irina Kiseleva and Natalie Larionova

isolated, huge progress has been made in studying, preventing, and treating influenza. Various aspects were discovered in detail, including the replication cycle of influenza viruses, genome structure [3, 4], ecology [5 - 7] and evolution [8 - 10], pathogenesis [11], mechanisms of the immune response to influenza virus infection [12], epidemiology of influenza [4], its express diagnostics [4], clinical course, and complications [13]. Full genome sequencing and reassortment by reverse genetics became routine techniques [14]. Many vaccines of new generation and effective antivirals were discovered [13, 15]. Health care programs on treatment and prevention of seasonal and pandemic influenza for clinical practice regarding the use of antiviral medications and vaccines were developed [16 - 19]. A chronology of key events in the flu research field can be seen in Table **1**.

Table 1. Timeline of selected key scientific discoveries in the influenza research field.

Year	Scientific Discovery or Event	Refs.
1918	The Spanish H1N1 influenza pandemic of 1918 spread quickly.	[20, 21]
1931	The first swine influenza A virus was isolated.	[22]
1933	The first human influenza A virus was isolated.	[23]
1937	The first mouse-derived inactivated influenza vaccine was developed.	[24]
1940	The first human influenza B virus was isolated.	[25]
1931	Embryonated chicken eggs were found to be a suitable substrate for virus isolation.	[26]
1938	The first mouse-derived live-attenuated influenza vaccine was developed by sequential passing of the virus through mice.	[27]
1942	Hemagglutination assay and hemagglutinin inhibition assay were developed for quantifying the concentration of viruses and antibodies.	[28]
1945	Mass scale immunization of healthy adults with influenza vaccines began.	[29, 30]
1947	The first human influenza C virus was isolated.	[31]
1957	The Asian H2N2 influenza pandemic of 1957 spread quickly	[20, 21]
1957	Susceptibility of primary cultures of embryonic chicken kidney cells was discovered.	[32]
1965	A new class of agents – adamantanes – was discovered; amantadine and rimantadine were shown to be effective against influenza A virus.	[33 - 35]
1965-1967	The first egg-derived cold-adapted live attenuated influenza vaccine was developed by sequential passing of the virus in embryonated chicken eggs at the low temperature of 25-26°C.	[36, 37]
1968	The Hong Kong H3N2 influenza pandemic of 1968 spread quickly.	[20, 21]
1977	The Russian H1N1 influenza pseudo-pandemic of 1977 spread quickly.	[21]
1978	WHO suggested that continuous MDCK cells provide a satisfactory alternative to eggs and monkey kidney cultures for the isolation of influenza viruses from clinical specimens.	[38]

(Table 1) cont.....

Year	Scientific Discovery or Event	Refs.
1985	Polymerase chain reaction (PCR) procedure was developed.	[39]
1997	The interspecies barrier was crossed (H5N1 outbreak in Hong Kong in 1997).	[40]
1999	Twice a year, WHO has issued annual recommendations for influenza vaccine formulations for the North and the South hemisphere.	[41]
2000	Neuraminidase inhibitors were developed.	[42]
2000	An eight-plasmid DNA transfection system for the rescue of influenza virus from cloned cDNA was developed.	[43]
2002	Cap-dependent endonuclease inhibitors were developed (PB1).	[44]
2005	Reconstruction of 1918 Spanish influenza pandemic virus.	[45]
2006	The architecture of RNP complexes in influenza A virus particles was discovered.	[46, 47]
2009	The 2009 H1N1 pandemic of 2009 spread quickly.	[20, 21]
2011	The first swine influenza D virus was isolated.	[48]
2012	The first WHO recommendations for the composition of a quadrivalent influenza vaccine were made.	[49]
2012	A militarized H5N1 avian influenza virus capable of easily transmitting among mammals was developed by reverse genetics. According to the authors, "it was the most dangerous virus that has ever been created.".	[50]
2012	United States FDA announced the approval of the first seasonal influenza cell-based inactivated vaccine.	[51]
2018	Cap-dependent endonuclease inhibitors were developed (PB2).	[52]
2018	It was discovered that just breathing is enough for a person with influenza to spread the virus; coughing or sneezing are not required.	[53]
2019	The structure and function of the influenza virus transcription and replication machinery were discovered in detail.	[3]

Advances in Physiology and Medicine

The Nobel Prize in physiology or medicine has been awarded 111 times between 1901 and 2020 [54]. Ten of them (9%) were granted to virologists for their tremendous success in the study of human papillomavirus, human immunodeficiency virus, prions, retroviral oncogenes, tumor-inducing viruses, poliomyelitis virus, a virus of yellow fever, and hepatitis C virus (Table **2**). Despite influenza being one of the most mysterious diseases inflicting our civilization for centuries, no Nobel Prize has been awarded for discoveries regarding influenza and how to combat it.

Is it Possible to Eradicate Influenza? the Paradox of the Current Situation

The paradox of the current situation is a clear contradiction between the achievements of theoretical science and practical results in the fight against influenza infection. Science has created all theoretical prerequisites for defeating influenza. Nevertheless, the study on influenza and influenza viruses is still at its beginning and the influenza virus is always one step ahead of us.

Having analyzed influenza viral ecology and relationships between animal and human influenza viruses, Ozawa and Kawaoka [7] suggested that "complete control of influenza viruses seems impossible". It seems that it is not futile to think that complete influenza eradication is possible in the nearest future.

Table 2. The Nobel Laureates in physiology or medicine awarded for their discoveries in the field of virology.

Year	Nobel Laureates	Breakthrough	Viruses
1951	Theiler M	Discoveries concerning yellow fever and how to combat it	Yellow fever virus
1954	Weller TH Robbins FC	Discovery of the ability of poliomyelitis viruses to grow in cultures of various types of tissue	Polioviruses
1966	Rous P	Discovery of tumor-inducing viruses	Oncoviruses
1969	Delbrück M Hershey AD Luria SE	Discoveries concerning the replication mechanism and the genetic structure of viruses	Bacteriophages
1975	Baltimore D Dulbecco R Temin HM	Discoveries concerning the interaction between tumor viruses and the genetic material of the cell	Oncoviruses
1989	Varmus HE	Discovery of the cellular origin of retroviral oncogenes	Retroviruses
1997	Prusiner SB	Discovery of Prions – a new biological principle of infection	Prions
2008	Barré-Sinoussi F Montagnier L	Discovery of the human immunodeficiency virus (HIV)	HIV
2008	zur Hausen H	Discovery of human papillomaviruses causing cervical cancer	Papillomaviruses
2020	Alter HJ Houghton M Rice CM	Discovery of the hepatitis C virus	Hepatitis C virus

CONCLUDING REMARKS

Nevertheless, we believe that science will eventually defeat human influenza one day. As we mentioned before, unfortunately, no Nobel Prize has been awarded for findings in influenza research until now. It is possible that those scientists who defeat influenza will become Nobel Laureates for their development of a universal remedy against influenza.

Another option for further developments of the epidemiological situation is also possible. It was noticed in multiple publications that reported rates of influenza have fallen significantly during coronavirus lockdowns [55 - 58]. The pandemic is probably developing according to the first rather than the second scenario (see Chapter **6**, Fig. **1**). The problem of influenza could be solved by itself, since influenza virus will be replaced with SARS-CoV-2. The future will show.

REFERENCES

[1] Johnson NP, Mueller J. Updating the accounts: global mortality of the 1918-1920 "Spanish" influenza pandemic. Bull Hist Med 2002; 76(1): 105-15. [http://dx.doi.org/10.1353/bhm.2002.0022] [PMID: 11875246]

[2] Starko KM. Salicylates and pandemic influenza mortality, 1918-1919 pharmacology, pathology, and historic evidence. Clin Infect Dis 2009; 49(9): 1405-10. [http://dx.doi.org/10.1086/606060] [PMID: 19788357]

[3] Fodor E, te Velthuis AJW. Structure and function of the influenza virus transcription and replication machinery. Cold Spring Harb Perspect Med 2020; 10(9): a038398. [http://dx.doi.org/10.1101/cshperspect.a038398] [PMID: 31871230]

[4] Krammer F, Smith GJD, Fouchier RAM, *et al.* Influenza. Nat Rev Dis Primers 2018; 4(1): 3. [http://dx.doi.org/10.1038/s41572-018-0002-y] [PMID: 29955068]

[5] Wille M, Holmes EC. The ecology and evolution of influenza viruses. Cold Spring Harb Perspect Med 2020; 10(7): a038489. [http://dx.doi.org/10.1101/cshperspect.a038489] [PMID: 31871237]

[6] Yoon SW, Webby RJ, Webster RG. Evolution and ecology of influenza A viruses. Curr Top Microbiol Immunol 2014; 385: 359-75. [http://dx.doi.org/10.1007/82_2014_396] [PMID: 24990620]

[7] Ozawa M, Kawaoka Y. Cross talk between animal and human influenza viruses. Annu Rev Anim Biosci 2013; 1: 21-42. [http://dx.doi.org/10.1146/annurev-animal-031412-103733] [PMID: 25387011]

[8] Tarro G. Origin of the viruses and their evolutionary history. Advanced Studies in Biology 2013; 5: 423. [http://dx.doi.org/10.12988/asb.2013.3835]

[9] Martin PM, Martin-Granel E. 2,500-year evolution of the term epidemic. Emerg Infect Dis 2006; 12(6): 976-80. [http://dx.doi.org/10.3201/eid1206.051263] [PMID: 16707055]

[10] Petrova VN, Russell CA. The evolution of seasonal influenza viruses. Nat Rev Microbiol 2018; 16(1): 47-60. [http://dx.doi.org/10.1038/nrmicro.2017.118] [PMID: 29081496]

[11] Qi L, Davis AS, Jagger BW, *et al.* Analysis by single-gene reassortment demonstrates that the 1918 influenza virus is functionally compatible with a low-pathogenicity avian influenza virus in mice. J Virol 2012; 86(17): 9211-20. [http://dx.doi.org/10.1128/JVI.00887-12] [PMID: 22718825]

[12] Oltz EM. Immunity to influenza: Closing in on a moving target. J Immunol 2019; 202(2): 325-6.

[http://dx.doi.org/10.4049/jimmunol.1890024] [PMID: 30617112]

[13] Beigel JH, Hayden FG. Influenza therapeutics in clinical practice-challenges and recent advances. Cold Spring Harb Perspect Med 2020; 11(4): a038463.
[http://dx.doi.org/10.1101/cshperspect.a038463] [PMID: 32041763]

[14] Blanco-Lobo P, Nogales A, Rodríguez L, Martínez-Sobrido L. Novel approaches for the development of live attenuated influenza vaccines. Viruses 2019; 11(2): E190.
[http://dx.doi.org/10.3390/v11020190] [PMID: 30813325]

[15] Biswas A, Chakrabarti AK, Dutta S. Current challenges: from the path of "original antigenic sin" towards the development of universal flu vaccines. Int Rev Immunol 2020; 39(1): 21-36.
[http://dx.doi.org/10.1080/08830185.2019.1685990] [PMID: 31707873]

[16] CDC. Influenza antiviral medications: Summary for clinicians. 2020. [Available from: https://www.cdc.gov/flu/professionals/antivirals/summary-clinicians.htm]

[17] CDC. Prevention and control of seasonal influenza with vaccines: Recommendations of the Advisory Committee on immunization practices — United States, 2019–20 Influenza Season. [Available from: https://www.cdc.gov/mmwr/volumes/68/rr/rr6803a1.htm]

[18] WHO. Global influenza preparedness plan. 2005. [Available from: http://www.who.int/csr/resources/publications/influenza/WHO_CDS_CSR_GIP_2005_5.pdf]

[19] WHO. Global action plan to increase vaccine supply for influenza vaccines. 2006. [Available from: http://whqlibdoc.who.int/hq/2006/WHO_IVB_06.13_eng.pdf]

[20] Flahault A, Zylberman P. Influenza pandemics: past, present and future challenges. Public Health Rev 2010; 32: 319-40.
[http://dx.doi.org/10.1007/BF03391605] [PMID: 32226192]

[21] Lina B. History of influenza pandemics. In: Raoult D, Drancourt M, Eds. Paleomicrobiology: Past Human Infections. Verlag Berlin Heidelberg: Springer 2008; pp. 199-211.
[http://dx.doi.org/10.1007/978-3-540-75855-6_12]

[22] Van Epps HL. Influenza: exposing the true killer. J Exp Med 2006; 203(4): 803.
[http://dx.doi.org/10.1084/jem.2034fta] [PMID: 16685764]

[23] Smith W, Andrewes CH, Laidlaw PP. A virus obtained from influenza patients. Lancet 1933; 222(5732): 66-8.
[http://dx.doi.org/10.1016/S0140-6736(00)78541-2]

[24] Francis T, Jr, Magill TP. The antibody response of human subjects vaccinated with the virus of human influenza. J Exp Med 1937; 65(2): 251-9.
[http://dx.doi.org/10.1084/jem.65.2.251] [PMID: 19870599]

[25] Francis T, Jr. A new type of virus from epidemic influenza. Science 1940; 92(2392): 405-8.
[http://dx.doi.org/10.1126/science.92.2392.405] [PMID: 17794275]

[26] Goodpasture EW, Woodruff AM, Buddingh GJ. The cultivation of vaccine and other viruses in the chorioallantoic membrane of chick embryos. Science 1931; 74(1919): 371-2.
[http://dx.doi.org/10.1126/science.74.1919.371] [PMID: 17810781]

[27] Zhdanov VM, Aleksandrova GI, Gendon IuZ. Live influenza vaccine in the USSR: its development, study and practical use. Zh Mikrobiol Epidemiol Immunobiol 1986; 7(7): 3-13.
[PMID: 3529762]

[28] Hirst GK. The quantitative determination of influenza virus and antibodies by means of red cell agglutination. J Exp Med 1942; 75(1): 49-64.
[http://dx.doi.org/10.1084/jem.75.1.49] [PMID: 19871167]

[29] Barberis I, Myles P, Ault SK, Bragazzi NL, Martini M. History and evolution of influenza control through vaccination: from the first monovalent vaccine to universal vaccines. J Prev Med Hyg 2016; 57(3): E115-20.

[PMID: 27980374]

[30] Crovari P, Alberti M, Alicino C. History and evolution of influenza vaccines. J Prev Med Hyg 2011; 52(3): 91-4.
[PMID: 22010533]

[31] Taylor RM. Studies on survival of influenza virus between epidemics and antigenic variants of the virus. Am J Public Health Nations Health 1949; 39(2): 171-8.
[http://dx.doi.org/10.2105/AJPH.39.2.171] [PMID: 18111593]

[32] Maassab HF. The propagation of multiple viruses in chick kidney cultures. Proc Natl Acad Sci USA 1959; 45(7): 1035-9.
[http://dx.doi.org/10.1073/pnas.45.7.1035] [PMID: 16590474]

[33] Cochran KW, Maassab HF, Tsunoda A, Berlin BS. Studies on the antiviral activity of amantadine hydrochloride. Ann N Y Acad Sci 1965; 130(1): 432-9.
[http://dx.doi.org/10.1111/j.1749-6632.1965.tb12579.x] [PMID: 4285587]

[34] Prichard WW, Hockessin D. Adamantanes and tricyclo[4.3.1.13.8] undecanes. US patent 3352912, 1967. [Available from: https://www.freepatentsonline.com/3352912.pdf]

[35] Rabinovich S, Baldini JT, Bannister R. Treatment of influenza. The therapeutic efficacy of rimantadine HC1 in a naturally occurring influenza A2 outbreak. Am J Med Sci 1969; 257(5): 328-35.
[http://dx.doi.org/10.1097/00000441-196905000-00005] [PMID: 4891873]

[36] Alexandrova GI, Smorodintsev AA. Obtaining of an additionally attenuated vaccinating cryophilic influenza strain. Rev Roum Inframicrobiol 1965; 2: 179-89.

[37] Maassab HF. Adaptation and growth characteristics of influenza virus at 25 degrees c. Nature 1967; 213(5076): 612-4.
[http://dx.doi.org/10.1038/213612a0] [PMID: 6040602]

[38] Davies HW, Appleyard G, Cunningham P, Pereira MS. The use of a continuous cell line for the isolation of influenza viruses. Bull World Health Organ 1978; 56(6): 991-3.
[PMID: 310738]

[39] Saiki RK, Scharf S, Faloona F, *et al.* Enzymatic amplification of beta-globin genomic sequences and restriction site analysis for diagnosis of sickle cell anemia. Science 1985; 230(4732): 1350-4.
[http://dx.doi.org/10.1126/science.2999980] [PMID: 2999980]

[40] CDC. Isolation of avian influenza A(H5N1) viruses from humans – Hong Kong, May–December 1997. MMWR 1997; 46: 12047. [Available from: https://www.cdc.gov/mmwr/preview/mmwrhtml/00050459.htm]

[41] WHO. Recommendations on the composition of influenza virus vaccines. 2020. [Available from: https://www.who.int/influenza/vaccines/virus/recommendations/en/]

[42] Gubareva LV, Kaiser L, Hayden FG. Influenza virus neuraminidase inhibitors. Lancet 2000; 355(9206): 827-35.
[http://dx.doi.org/10.1016/S0140-6736(99)11433-8] [PMID: 10711940]

[43] Hoffmann E, Neumann G, Kawaoka Y, Hobom G, Webster RG. A DNA transfection system for generation of influenza A virus from eight plasmids. Proc Natl Acad Sci USA 2000; 97(11): 6108-13.
[http://dx.doi.org/10.1073/pnas.100133697] [PMID: 10801978]

[44] Furuta Y, Takahashi K, Fukuda Y, *et al.* *In vitro* and *in vivo* activities of anti-influenza virus compound T-705. Antimicrob Agents Chemother 2002; 46(4): 977-81.
[http://dx.doi.org/10.1128/AAC.46.4.977-981.2002] [PMID: 11897578]

[45] Tumpey TM, Basler CF, Aguilar PV, *et al.* Characterization of the reconstructed 1918 Spanish influenza pandemic virus. Science 2005; 310(5745): 77-80.
[http://dx.doi.org/10.1126/science.1119392] [PMID: 16210530]

[46] Noda T, Sagara H, Yen A, *et al.* Architecture of ribonucleoprotein complexes in influenza A virus

particles. Nature 2006; 439(7075): 490-2.
[http://dx.doi.org/10.1038/nature04378] [PMID: 16437116]

[47] Zheng W, Tao YJ. Structure and assembly of the influenza A virus ribonucleoprotein complex. FEBS Lett 2013; 587(8): 1206-14.
[http://dx.doi.org/10.1016/j.febslet.2013.02.048] [PMID: 23499938]

[48] Hause BM, Ducatez M, Collin EA, *et al.* Isolation of a novel swine influenza virus from Oklahoma in 2011 which is distantly related to human influenza C viruses. PLoS Pathog 2013; 9(2): e1003176.
[http://dx.doi.org/10.1371/journal.ppat.1003176] [PMID: 23408893]

[49] WHO. Recommended composition of influenza virus vaccines for use in the 2013 Southern hemisphere influenza season. 2012. [Available from: http://www.who.int/influenza/vaccines/virus/recommendations/2013_south/en/.17]

[50] Herfst S, Schrauwen EJ, Linster M, *et al.* Airborne transmission of influenza A/H5N1 virus between ferrets. Science 2012; 336(6088): 1534-41.
[http://dx.doi.org/10.1126/science.1213362] [PMID: 22723413]

[51] Moro PL, Winiecki S, Lewis P, Shimabukuro TT, Cano M. Surveillance of adverse events after the first trivalent inactivated influenza vaccine produced in mammalian cell culture (Flucelvax(®)) reported to the Vaccine Adverse Event Reporting System (VAERS), United States, 2013-2015. Vaccine 2015; 33(48): 6684-8.
[http://dx.doi.org/10.1016/j.vaccine.2015.10.084] [PMID: 26518405]

[52] Pflug A, Gaudon S, Resa-Infante P, *et al.* Capped RNA primer binding to influenza polymerase and implications for the mechanism of cap-binding inhibitors. Nucleic Acids Res 2018; 46(2): 956-71.
[http://dx.doi.org/10.1093/nar/gkx1210] [PMID: 29202182]

[53] Yan J, Grantham M, Pantelic J, *et al.* Infectious virus in exhaled breath of symptomatic seasonal influenza cases from a college community. Proc Natl Acad Sci USA 2018; 115-1086.(5): 1081-6.
[http://dx.doi.org/10.1073/pnas.1716561115] [PMID: 29348203]

[54] The Nobel Prize in Physiology or Medicine. 2020. [Available from: https://www.nobelprize.org/prizes/lists/all-nobel-laureates-in-physiology-or-medicine/].

[55] Chan KH, Lee P-W, Chan CY, Lam KBH, Ho P-L. Monitoring respiratory infections in covid-19 epidemics. BMJ 2020; 369: m1628.
[http://dx.doi.org/10.1136/bmj.m1628] [PMID: 32366507]

[56] Jones N. How coronavirus lockdowns stopped flu in its tracks. Nature 21 May 2020.
[http://dx.doi.org/10.1038/d41586-020-01538-8] [PMID: 32440021]

[57] Sullivan SG, Carlson S, Cheng AC, *et al.* Where has all the influenza gone? The impact of COVID-19 on the circulation of influenza and other respiratory viruses, Australia, March to September 2020. Euro Surveill 2020; 25(47): 2001847.
[http://dx.doi.org/10.2807/1560-7917.ES.2020.25.47.2001847] [PMID: 33243355]

[58] Nowak MD, Sordillo EM, Gitman MR, Paniz Mondolfi AE. Coinfection in SARS-CoV-2 infected patients: Where are influenza virus and rhinovirus/enterovirus? J Med Virol 2020; 92(10): 1699-700.
[http://dx.doi.org/10.1002/jmv.25953] [PMID: 32352574]

Abbreviations

3Rs Replacement, reduction, and refinement

A/AA A/Ann Arbor/6/60ca

ARD Acute respiratory disease

att Attenuated phenotype

B/AA B/Ann Arbor/1/66ca

B60 B/USSR/60/69 MDV

BCG Bacille Calmette-Guérin vaccine

ca Cold-adapted phenotype

CDC Centers for Disease Control and Prevention, the US

CE Developing chicken embryos

cRNA Complementary RNA

DNA Deoxyribonucleic acid

EID_{50} 50% chicken embryo infectious dose

ELISA Enzyme-linked immunosorbent assay

FDA Food and Drug Administration, the US

Flu Influenza

H3N2v Swine H3N2 variant viruses with the M gene from the H1N1pdm09

HA Hemagglutination assay; Hemagglutinin

HAI Inhibition of hemagglutination

HEF Hemagglutinin-esterase-fusion protein

HIV Human immunodeficiency virus

HPAI Highly pathogenic avian influenza

IEM Institute of Experimental Medicine, Russia

IFA Immunofluorescence antibody staining

IIV Inactivated influenza vaccine

L17 A/Leningrad/134/17/57 MDV

L47 A/Leningrad/134/47/57 MDV

LAIV Live attenuated influenza vaccine

LD_{50} 50% lethal dose

LPAI Low pathogenic avian influenza

LRT Low respiratory tract

LWT A/Leningrad/134/57, the WT progenitor of L17 MDV

MDCK Madin-Darby canine kidney cells
MDV Master donor virus
MID_{50} 50% minimum infectious dose
MMR Measles, mumps, and rubella vaccine
MN Microneutralisation
NA Neuraminidase
NA Neuraminidase assay
NAI Neuraminidase inhibitor assay
NEP Nuclear export protein
NP Nucleoprotein
PAAG Polyacrylamide gel
PAGE Polyacrylamide gel electrophoresis
PCK Primary chick kidney cells
PFU Plaque-forming units
PR8 A/Puerto Rico/8/34
RCT Reproductive capacity at different temperatures
RIDT Rapid influenza diagnostic test
RNA Ribonucleic acid
RNP Ribonucleoprotein
RS-virus Respiratory syncytial virus
RT-PCR Reverse transcription polymerase chain reaction
SGR Single-gene reassortant
$TCID_{50}$ 50% tissue culture infectious dose
ts Temperature sensitive phenotype
URT Upper respiratory tract
US United States
USSR Union of Soviet Socialist Republics
VLPs Virus-like particle
vRNA Viral ribonucleic acid
vRNP Viral ribonucleoprotein
WHO World Health Organization
WT Wild-type

SUBJECT INDEX

A

B

Irina Kiseleva and Natalie Larionova

E

F

G

H

I

K

M

N

O

P

R

S

T

V

Made in the USA
Las Vegas, NV
03 November 2026

57422842R00128